T0272614

First published in 2011 by New Society Publishers

This reprint published in 2024 by Sphinx Books, an imprint of Aeon Books

British Library Cataloguing in Publication Data

A C.I.P. for this book is available from the British Library
ISBN-13: 9781915952233

Cover Art by Phoebe Young

THE WEALTH
of NATURE

THE WEALTH
of NATURE

· · · · ·

ECONOMICS *as if*
SURVIVAL MATTERED

John Michael Greer

SPHINX

Contents

Introduction:
A Guide for the Perplexed

More than two centuries have passed since Adam Smith, a Scots philosopher with a clergyman's training and a previous reputation mostly as a moralist, launched the modern science of economics with a book entitled *The Wealth of Nations*. The first widely accepted analysis of how markets guide economic behavior, Smith's book quickly took on the status of a classic. Its approach to the subject of economics has dominated the field ever since; just as Bertrand Russell famously defined all of Western philosophy as "footnotes to Plato," it would be by no means inappropriate to define all of modern economic thought as footnotes to Smith.

While Smith's work deserves its reputation, the timing of its publication also had a good deal to do with the way it came to dominate the field it brought into being. It was published the same year that America's Declaration of Independence was signed, and just as Britain was launching the social and technological transformations that would make it the world's first industrial society. Smith's central thesis — that a market economy governs itself through what would today be called feedback loops, and thus does not need close government supervision — appealed to audiences who were well primed by the events of the day to question the need for government intrusion in private economic affairs, while the rising class of industrial entrepreneurs found their own reasons to support an analysis that claimed to justify their interest in making as much money with as little interference as possible.

The meteoric expansion of the industrial economy over the two centuries that followed was marked, in turn, by the rise to prominence of the profession of economics. In Smith's time, professional economists did not yet exist; today they rank among the most prominent figures in contemporary intellectual life. The creation of a new Nobel prize in economics in 1968[1] can be taken as a suitable marker of the profession's coming of age. Still, there is a discordant note in all this, for the rise of economics as a science and a profession has not been accompanied by any noticeable improvement in the ability of societies to manage their economic affairs.

It's ironic , in fact, how few benefits industrial societies seem to have gained from their economic experts in the last few decades. Beginning around 1980, when many of the world's industrial nations adopted a market-centered economic philosophy based on recent revisions of Adam Smith's ideas, it has become routine for government policies to have results completely different from, and far less pleasant than, those predicted for these policies by public and private economists alike. From the deregulation of utilities in the early 1980s, which was supposed to cause prices of utility services to go down (and made those prices go up), through a litany of failures culminating in the drastic cuts in interest rates that followed the 1999–2000 tech stock crash, which was supposed to restore economic stability (and launched an even more drastic cycle of boom and bust), it has become uncomfortably clear that whatever the talents of today's economic profession happen to be, making meaningful predictions about economic policy is not one of them.

There has been plenty of discussion of these failures, to be sure, and a great many attempts to tinker with the structure of contemporary economic thought; many writers in and out of the profession have attempted to explain the failures and prevent their recurrence. Still, as this book will try to show, most of these efforts stop short of the changes that will actually be needed to bring economics back in line with reality. The problems with contemporary economics cannot be fixed by minor adjustments to existing models. They reach down to the basic assumptions on which Adam Smith and his successors based the science of economics, and can be fixed only by recognizing the flaws in those assumptions and exchanging them for others more in keeping with the way the world actually works.

It is not going too far to compare Smith to Claudius Ptolemy, the great classical astronomer whose writings became the foundation for all cosmological and astronomical thought for a millennium after his time. Ptolemy's theories were clear and compelling, and they corresponded closely enough to the way the heavens seemed to work that most of the astronomers who built on his work took his models for hard fact. Underlying those models, though, was a fundamental mistake: Ptolemy's conviction that the Earth was located at the center of the cosmos and everything else rotated around it. That mistake forced Ptolemy's successors to come up with one workaround after another for celestial motions that never quite did what the model said they should do, until Nicolaus Copernicus eventually figured out where the problem actually was. By turning the cosmos inside out, moving the supposedly central Earth to a peripheral position and putting the Sun at the center, he solved problems that could not be solved from within Ptolemy's analysis, and also opened up avenues of inquiry that made possible Kepler's discovery of the laws of planetary motion, Newton's theory of gravitation and a great deal more.

Economics is badly in need of a Copernican revolution of its own, one that will recognize that the center of economic activity is not where today's economists think it is. Fortunately, the field has already had its Copernicus.

Ernst Friedrich Schumacher was born in Bonn in 1911 and attended universities there and in Berlin before going to Oxford in 1930 as a Rhodes Scholar, and then to Columbia University in New York, where he graduated with a doctorate in economics. When the Second World War broke out he was living in Britain, and was interned for a time as an enemy alien, until fellow economist John Maynard Keynes arranged for his release. After the war, he worked for the British Control Commission, helping to rebuild the West German economy, and then began a 20-year stint as chief economist and head of planning for the British National Coal Board, which at the time was one of the world's largest energy firms.

He also served as an economic adviser to the governments of India, Burma and Zambia, and these experiences turned his attention to the economic challenges of development in the Third World. Recognizing that attempts to import the industrial model into nonindustrial countries usually failed due to shortages of infrastructure and resources, he pioneered the concept of intermediate technology — an approach to development that focuses on finding and using the technology best suited to the resources available — and founded the Intermediate Technology Development Group in 1966. His interest in resource issues also led to an involvement in the organic agriculture movement, and he served for many years as a director of the Soil Association, Britain's largest organic farming organization.

It was these practical involvements that predisposed Schumacher to see past the haze of unrecognized ideology that makes so much contemporary economic thought useless when applied to the real world. The academic side of the economics profession is notoriously forgiving of even the most embarrassingly inaccurate predictions, and a professor of economics can still count on being taken seriously even when every single public statement he has made about future economic conditions has been utterly disproven by events. This is much less true in the business world, where predictions have results measured in quarterly profits or losses. Working in a setting where consistently bad predictions would have cost him his job, Schumacher was not at liberty to put theory ahead of evidence, and the conflict between what standard economic theory said and the realities Schumacher observed all around him must have had a role in making him the foremost economic heretic of his time.

His economic ideas cover a great deal of ground, not all of which could be explored even in a book many times the length of this one. Four of his propositions, however, will help sketch out the gap he discovered between economic theory and the behavior of economic systems in the real world.

First, Schumacher drew a hard distinction between *primary goods* and *secondary goods*. The latter term includes most of what is dealt with by conventional economics: the goods and services produced by human labor and exchanged among human beings. The former includes all those things necessary for human life and economic activity that are produced not by human beings, but by Nature. Schumacher pointed out that primary goods need to come first in any economic analysis, because they supply the preconditions for the production of secondary goods. Renewable resources, he proposed, form the equivalent of income in the primary economy, while nonrenewable resources are the equivalent of capital; to insist that an economic system is sound when it is burning through nonrenewable resources at a rate that will lead to rapid depletion is thus as silly as claiming that a business is breaking even if it's covering up huge losses by drawing down its bank accounts.

Second, Schumacher stressed the central role of energy among primary goods. He argued that energy cannot be treated as one commodity among many without reducing economics to gibberish, because energy is the gateway resource that gives access to all other resources. Given enough energy, shortages of any other resource can be made good one way or another; if energy runs short, though, abundant supplies of other resources won't make up the difference, because any economy needs energy to bring those resources into the realm of secondary goods and make them available for human needs. Thus the amount of energy available per person puts an upper limit on the level of economic development possible in a society, though other forms of development—social, intellectual, spiritual—can still be pursued in a setting where hard limits on energy restrict economic life.

Third, Schumacher stressed the importance of a variable left out of most economic analyses—the cost per worker of establishing and maintaining a workplace. Only the abundant capital, ample energy supplies and established infrastructure of the world's industrial nations, he argued, made it possible for businesses and governments in those nations to treat replacing human labor with technology as a benefit. In the nonindustrial world, where the most urgent economic task was not the production of specialty goods for global markets but the provision of paid employment and basic necessities to the local population, attempts at industrialization have commonly turned into costly mistakes. Schumacher's involvement in intermediate technology unfolded from this realization; he pointed out that in a great many situations, a relatively simple technology that relied on human hands and minds to meet local needs with local resources was the most viable response to the economic needs of nonindustrial nations.

Finally, and most centrally, Schumacher pointed out that the failures of contemporary economics could not be solved by improved mathematical models or more detailed statistics, because they were hardwired into the assumptions underlying economics itself. Every way of thinking about the world rests ultimately on presuppositions that are, strictly speaking, metaphysical in Nature: that is, they deal with fundamental questions about what exists and what has value. Trying to ignore the metaphysical dimension does not make it go away, but rather simply ensures that those who make this attempt will be blindsided whenever the real world fails to behave according to their unexamined assumptions. Contemporary economics fails to predict the behavior of the economy because it fails to criticize its own underlying metaphysics.

Thus, a hard look at those basic assumptions is an unavoidable part of straightening out the mess into which current economic ideas have helped land us.

These four propositions are among the key points made in Schumacher's most famous book, *Small Is Beautiful: Economics As If People Mattered*, which was published in 1973 to immediate acclaim. Theodore Roszak spoke for many perceptive thinkers at that time in the closing words of his introduction: "We need a nobler economics that is not afraid to discuss spirit and conscience, moral purpose and the meaning of life, an economics that aims to educate and elevate people, not merely to measure their low-grade behavior. Here it is."[2] The more thoughtful end of the alternative scene of the seventies took Schumacher's ideas as one of the core foundations of its thought and practice; it's no exaggeration to say that any bookshelf in that decade that had a copy of *The Whole Earth Catalog* and *Mother Earth News* on it generally had a copy of *Small Is Beautiful* perched nearby.

Like the rest of the alternative thinking of that decade, Schumacher's work went into eclipse in the following decades as industrial societies abandoned the promising steps toward sustainability taken in the seventies and embraced a radical new ideology that insisted on the free market's infallibility. For a while, that new ideology seemed to work, but its successes rested on a foundation rarely discussed, then or later. The seventies had been defined in part by repeated energy crises and resulting economic troubles. Those difficulties helped catalyze a sharp decrease in energy use in the world's industrial nations; it also drove frenzied efforts to locate untapped petroleum reserves — efforts that, as it happened, turned up large fields in the North Sea and Alaska's North Slope.

A reasonable approach to energy policy would have treated those discoveries as vital sources of energy to fill in the gaps while industrial nations made the transition to a new economy powered by renewable sources and structured to maximize energy efficiency—exactly the policy that Schumacher, among many others, detailed in his writings. Unfortunately for all of us, that was not what happened. Instead, the North Sea and North Slope fields were pumped at a breakneck pace, flooding oil markets around the world with cheap oil and sending prices plummeting to what, corrected for inflation, was their lowest level in history. As the efficiencies and innovative programs of the seventies gave way to the extravagances of the eighties and nineties, the possibility of a smooth transition to a sustainable future went by the boards.

We are now living with the first consequences of that monumentally short-sighted decision. The wild swings in energy costs, the even wilder cycles of economic boom and bust and the sharp impacts of all this volatility on the social and political fabric of the world's industrial nations have their source in the disastrous mismatch between an economic and technological system geared to exponential growth and the hard limits of a finite planet.[3] The consequences to our descendants will be even more extreme. Still, it may be possible to mitigate the worst of those consequences, and make life considerably easier for generations to come, by revisioning the ways that human societies deal with the production and distribution of goods and services. The crisis of the present makes such a revisioning necessary, but it also provides a window of opportunity in which such a revision might just be able to find its way from theory into practice.

This book is meant to further that hard but necessary process. I am not a professional economist, and the cult of expertise that pervades modern culture may make it seem presumptuous for someone without an economics degree to suggest a redefinition of contemporary economics. It may seem even more presumptuous in that several movements toward an ecologically sane economics have risen since Schumacher's time; economists such as Nicholas Georgescu-Roegen, Kenneth Boulding, Herman Daly and Robert Costanza have presented densely reasoned works of economic theory arguing for the inclusion of the value of "natural capital" — in Schumacher's terms, primary goods — in economic calculations, and challenging the simplistic ways that conventional economic thought brushes aside environmental destruction and resource depletion as nonissues.

Still, it may be worth noting that Adam Smith did not have an economics degree, and his pathbreaking study of economics lacked any trace of the intricate mathematical formulations that today's economists too often consider essential to their craft. Smith worked instead on the level of fundamental ideas; Schumacher, whose writings are just as sparse in their use of calculations, did the same; so, in its own way, does this book. Like Smith's and Schumacher's works, too, this book is not addressed to economists. Rather, its goal is to communicate the failure of modern economics, and potential solutions to the crises driven by that failure, to the audience that has the final say in matters of public policy: the public itself.

The predicament into which industrial civilization has backed itself at the end of the age of cheap energy, the subject of my previous books *The Long Descent* and *The Ecotechnic Future*, has a crucial economic dimension, and *The Wealth of Nature* is intended to help make sense of that dimension by applying E. F. Schumacher's Copernican revolution in economic thought to the crisis of our time. If the ideas suggested here inspire others to rethink the foundations of contemporary economics, and set aside some of the mistaken ideas that have so consistently blocked a sane response to the largely unacknowledged roots of our current troubles, this book will have done its work.

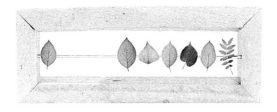

THE FAILURE *of* ECONOMICS

IMAGINE FOR A MOMENT that you are on board a sailing ship in the middle of the ocean. You wake in the middle of the night with an uneasy feeling, as if trouble is brewing. You get dressed and go on deck. It's a clear night with a steady wind, and you can see some distance over the water; as you glance off to starboard, though there is no land in sight, you are horrified to see waves crashing over black jagged rocks not far from the ship, setting the sea afoam.

You hurry aft to the midship bridge to warn the crew members on watch, and find the first mate and several other crew members sitting there, calmly smoking their pipes and paying no attention to the rocks. When you ask them about the rocks, they deny that any such thing exists in that part of the ocean, and insist that what you've seen is an optical illusion common in those latitudes. One of the crew members takes you into the chart room and shows you a chart with the ship's progress marked on it. Sure enough, there are no rocks anywhere near the ship's course, but as you glance over the chart you realize that there are no rocks marked anywhere else, either, nor any reefs, shoals or other hazards to navigation.

You leave the chart room, shaking your head, and glance at the compass in the binnacle. This only increases your discomfort; its needle indicates that magnetic north ought to be off the port bow, but a glance up at the sky shows the Little Dipper dead astern. When you mention this to the crew members, though, they roll their eyes and tell you that you obviously haven't studied navigation. You leave the midship bridge and walk forward, looking ahead to see where the ship is going, and sure enough, the pale gleam of rough water around rocks shows up in the distance.

It would be comforting if this scenario was just a nightmare; unfortunately, it mirrors one of the most troubling realities of contemporary life. The metaphoric charts and compass used nowadays to guide most of the important decisions made by the world's nations come from the science of economics, and the policy recommendations presented by economists to decision makers and ordinary people alike consistently fail to provide useful guidance in the face of some of the most central challenges of our time.

This may seem like an extreme statement, but the facts to back it up are as close as the nearest Internet news site. Consider the way that economists responded — or, rather, failed to respond — to the gargantuan multinational housing boom that imploded so spectacularly in 2008, taking much of the global economy with it.[1] This was as close to a perfect example of a runaway speculative bubble as you'll find anywhere in recent history. The extensive literature on speculative bubbles, going back all the way to Rev. Charles Mackay's *Extraordinary Popular Delusions and the Madness of Crowds*, made it no challenge at all to recognize that the housing boom was simply another example of this species. All the classic symptoms were present and accounted for: the dizzying price increases, the huge influx of amateur investors, the giddy rhetoric insisting that prices could and would keep on rising forever, the soaring rate of speculation using borrowed money and more.

By 2005, accordingly, a good many people outside the economics profession were commenting on parallels between the housing bubble and other speculative binges. By 2006 the blogosphere was abuzz with accurate predictions of the approaching crash, and by 2007 the final plunge into mass insolvency and depression was treated in many circles as a foregone conclusion — as indeed it was by that time. Keith Brand, who founded the lively Housing-Panic blog in 2005 to publicize the approaching disaster, and kept up a running stream of acerbic commentary straight through the bubble and bust, summarized those predictions with a tag line that could serve as the epitaph for the entire housing frenzy: "Dear God, this is going to end so badly." [2]

Yet it's a matter of public record that among those who issued these warnings, economists were as scarce as hen's teeth. Rather, most economists at the time dismissed the idea that the housing boom could be what it patently was, a disastrous speculative bubble. Nouriel Roubini, one of the few exceptions, has written wryly about the way he was dismissed as a crank for pointing out what should have been obvious to everybody else in his profession. [3] For whatever reason, it was not obvious at all; the vast majority of economists who expressed a public opinion on the bubble while it was inflating insisted that the delirious rise in real estate prices was justified, and that the exotic financial innovations that drove the bubble would keep banks and mortgage companies safe from harm.

These comforting announcements were wrong. Those who made them should have known, while the words were still in their mouths, that they were wrong. No less an economic luminary than John Kenneth Galbraith pointed out many decades ago that in the financial world, the term "innovation" inevitably refers to the rediscovery of the same small collection of emotionally appealing bad ideas that always lead to economic disaster when they are applied to the real world. [4] Galbraith's books *The Great Crash 1929* and *A Short History of Financial Euphoria*, which chronicle the repeated carnage caused by these same bad ideas in the past, can be found on the library shelves of every school of economics in North America, and anyone who reads either one can find every rhetorical excess and fiscal idiocy of the housing bubble faithfully duplicated in the great speculative binges of the past.

If the housing bubble were an isolated instance of failure on the part of the economics profession, it might be pardonable, but the same pattern of reassurance has repeated itself as regularly as speculative bubbles themselves. The same assurances were offered—in some cases, by the same economists—during the last great speculative binge in American economic life, the tech-stock bubble of 1996–2000. Identical assurances have been offered by the great majority of professional economists during every other speculative binge since Adam Smith's time. More than two hundred years of glaring mistakes would normally be considered an adequate basis for learning from one's errors, but in this case it has apparently been insufficient.

The Illusion of Invincibility

The problem with contemporary economics can be generalized as a blindness to potential disaster. This can be traced well outside the realm of bubble economics. Consider the self-destruction of Long Term Capital Management (LTCM) in 1998.[5] LTCM was one of the first high-profile hedge funds, and made money—for a while, quite a bit of it—by staking huge amounts of other people's funds on complex transactions based on intricate computer algorithms. It prided itself on having two Nobel laureates in economics on staff. Claims circulating on Wall Street during the firm's glory days had it that LTCM's computer models were so good that they could not lose money in the lifetime of this universe or three more like it.

Have you ever noticed that villains in bad science fiction movies usually get blown to kingdom come a few seconds after saying "I am invincible"? Apparently the same principle applies in economics, though the time lag is longer. It was some five years after LTCM launched its computer-driven strategy that the universe ended, slightly ahead of schedule. LTCM got blindsided by a Russian foreign-loan default that many other people saw coming, and failed catastrophically. The US government had to arrange a hurried rescue package to keep the implosion from causing a general financial panic.

Economists are not, by and large, stupid people. Many of them are extraordinarily talented; the level of mathematical skill displayed by the number-crunching "quants" in today's brokerages and investment banks routinely rivals that in leading university physics departments. Somehow, though, many of these extremely clever people have not managed to apply their intelligence to the task of learning from a sequence of glaring and highly publicized mistakes. This is troubling for any number of reasons, but the reason most relevant just now is that economists play a leading role among those who insist that industrial economies need not trouble themselves about the impact of limitless economic growth on the biosphere and the resource base that supports all our lives. If they turn out to be as wrong about that as so many economists were about the housing bubble, they will have made a fateful leap from risking billions of dollars to risking billions of lives.

Thus it's urgent to talk about the reasons why the economic mainstream has so often been unable to anticipate the downside. Like most of the oddities of contemporary life, this blindness to trouble has many causes. Two important ones result from peculiarities in the profession of economics as presently practiced; a third and more important reason is rooted in the fundamental assumptions that professional economists apply to the challenges of their field. The first two deserve discussion, but it's the third that will lead into the central project of this book: the quest for economic insights that will help make sense of the challenges of industrial society's future.

The first of the factors peculiar to the profession is that, for professional economists, *being wrong is usually much more lucrative than being right*. During the run-up to a speculative binge, and even more so during the binge itself, many people are willing to pay handsomely to be told that throwing their money into the speculation du jour is the right thing to do. Very few people are willing to pay to be told that they might as well flush their life's savings down the toilet, even — indeed, especially — when this is the case. During and after the crash, by contrast, most people have enough demands on their remaining money that paying economists to say anything at all is low on the priority list.

This rule applies to professorships at universities, positions at brokerages and many of the other sources of income open to economists. When markets are rising, those who encourage people to indulge their fantasies of overnight unearned wealth will be far more popular, and thus more employable, than those who warn them of the inevitable outcome of such fantasies; when markets are plunging, and the reverse might be true, nobody's hiring. Apply the same logic to the future of industrial society and the results are much the same: those who promote policies that allow people to get rich and live extravagantly today can count on an enthusiastic response, even if those same policies condemn industrial society to a death spiral in the decades ahead. Posterity pays nobody's salaries today.

The second of the forces driving bad economic advice is shared with many other contemporary fields of study: *economics suffers from a bad case of premature mathematization*. The dazzling achievements of the natural sciences have encouraged scholars in a great many fields to ape scientific methods in the hope of duplicating their successes, or at least cashing in on their prestige. Before Isaac Newton could make sense of planetary movements, though, thousands of observational astronomers had to amass the raw data with which he worked. The same thing is true of any successful science:

what used to be called "natural history," the systematic recording of what Nature actually does, builds the foundation on which later scientists erect structures of hypothesis and experiment.

Too many fields of study have attempted to skip these preliminaries and fling themselves straight into the creation of complex mathematical formulas, on the presumption that this is what real scientists do. The results have not been good, because there's a booby trap hidden inside the scientific method: the fact that you can get some fraction of Nature to behave in a certain way under arbitrary conditions in the artificial setting of a laboratory does not mean that Nature behaves that way when left to herself. If all you want to know is what you can force a given fraction of Nature to do, this is well and good, but if you want to understand how the world works, the fact that you can often force Nature to conform to your theory is not exactly helpful. Theories that are not checked against the evidence of observation reliably fail to predict events in the real world.

Economics is particularly vulnerable to the negative impact of premature mathematization because its raw material—the collective choices of human beings making economic decisions—involves so many variables that the only way to control them all is to impose conditions so arbitrary that the results have only the most distant relation to the real world. The logical way out of this trap is to concentrate on the equivalent of natural history, which is economic history: the record of what has actually happened in human societies under different economic conditions. This is exactly what those who predicted the housing crash did: they noted that a set of conditions in the past (a bubble) consistently led to a common result (a crash) and used that knowledge to make accurate predictions about the future.

Yet this is not, on the whole, what successful economists do nowadays. Instead, a great many of them spend their careers generating elaborate theories and quantitative models that are rarely tested against the evidence of economic history. The result is that when those theories are tested against the evidence of today's economic realities, they fail.

The Nobel laureates whose computer models brought LTCM crashing down in flames, for example, created what amounted to extremely complex hypotheses about economic behavior, and put those hypotheses to a very expensive test, which they failed. If they had taken the time to study economic history first, they might well have noticed that politically unstable countries often default on their debts, that moneymaking schemes involving huge amounts of other people's money normally implode and that every previous attempt to profit by modeling the market's vagaries had come to grief when confronted by the sheer cussedness of human beings making decisions about their money. They did not notice these things, and so they and their investors ended up losing astronomical amounts of money.

The inability of economics to produce meaningful predictions has become proverbial even within the profession. Even so mainstream an economic thinker as David A. Moss, a Harvard Business School professor and author of the widely quoted and utterly orthodox *A Concise Guide to Macroeconomics*, warns:

> Unfortunately, some students of macroeconomics are so confident about what they have learned that they refuse to see departures at all, preferring to believe that the economic relationships defined in their textbooks are inviolable rules. This sort of arrogance (or narrow-mindedness) becomes a true hazard to society when it infects macroeconomic policy making. The policy maker who believes he or she knows exactly how the economy will respond to a particular stimulus is a very dangerous policy maker indeed.[6]

Yet this understates the problem by a significant margin, because a great many of the pronouncements made these days by economists are not merely full of uncertainties; they are quite simply wrong. The quest to turn economics into a quantitative science in advance of the necessary data collection has produced far too many elegant theories that not only fail to model the real world, but consistently make inaccurate predictions. This would be bad enough if these theories were safely locked away in the ivory towers of academe; unfortunately this is far from the case nowadays. Much too often, theories that have no relation to the realities of economic life are used to guide business decisions and government policies, with disastrous results.

The Failure of Markets

The third force driving the economic profession's blindness to the downside is more complex than the two just discussed, because it deals not with the professional habits of economists but with the fundamental assumptions about the world that underlie economics as practiced today. Perhaps the most important of those is the belief in the infallibility of free markets. *The Wealth of Nations* popularized the idea that free market exchanges offered a more efficient way of managing economic activity than custom or government regulation. The popularity of Adam Smith's arguments on this subject has waxed and waned over the years; it may come as no surprise that periods of general prosperity have seen the market's alleged wisdom proclaimed to the skies, while periods of depression and impoverishment have had the reverse effect.

The economic orthodoxy in place in the Western world since the 1950s, neoclassical economics, has made a nuanced version of Smith's theory central to its approach to market phenomena. Neoclassical economists argue that, aside from certain exceptions discussed in the technical literature, people make rational decisions to maximize benefits to themselves, and the sum total of these decisions maximizes the benefits to everyone. The concept of market failure is part of the neoclassical vocabulary, and some useful work has been done under the neoclassical umbrella to explain how it is that markets can fail to respond to crucial human needs, as they so often do. Still, as already pointed out, neoclassical economists have consistently failed to foresee the most devastating examples of market failure, the speculative booms and busts that have rocked the global economy to its foundations, or even to recognize them while they were happening.

This is not the only repeated failure that can be chalked up to the discredit of the neoclassical consensus. Social critics have commented, for example, on the ease with which neoclassical economics ignores the interface between economic wealth and power. Even when people rationally seek to maximize benefits to themselves, after all, their options for doing so are very often tightly constrained by economic systems that have been manipulated to maximize the benefits going to someone else.

This is a pervasive problem in most human societies, and it's worth noting that those societies that survive over the long term tend to be the ones that work out ways to keep too much wealth from piling up uselessly in the hands of those with more power than others. This is why hunter-gatherers have customary rules for sharing out the meat from a large kill, why traditional mores in so many tribal societies force chieftains to maintain their positions of influence by lavish generosity and why those nations that got through the last Great Depression intact did so by imposing sensible checks and balances on concentrated wealth.

By neglecting and even arguing against these necessary redistributive processes, neoclassical economics has helped feed economic disparities, and these in turn have played a major role in driving cycles of boom and bust. It's not an accident that the most devastating speculative bubbles happen in places and times when the distribution of wealth is unusually lopsided, as it was in America, for example, in the 1920s and the period from 1990 to 2008. When wealth is widely distributed, more of it circulates in the productive economy of wages and consumer purchases; when wealth is concentrated in the hands of a few, more of it moves into the investment economy where the well-to-do keep their wealth, and a buildup of capital in the investment economy is one of the necessary preconditions for a speculative binge.

At the same time, concentrations of wealth can be cashed in for political influence, and political influence can be used to limit the economic choices available to others. Individuals can and do rationally choose to maximize the benefits available to them by exercising influence in this way, and the results impose destructive inefficiencies on the whole economy; the result is one type of market failure. In effect, political manipulation of the economy by the rich for private gain does an end run around normal economic processes by way of the world of politics; what starts in the economic sphere, as a concentration of wealth, and ends there, as a distortion of the economic opportunities available to others, ducks through the political sphere in between.

A similar end run drives speculative bubbles, although here the non-economic sphere involved is that of crowd psychology rather than politics. Very often, the choices made by participants in a bubble are not rational decisions that weigh costs against benefits. A speculative bubble starts in the economic sphere as a buildup of excessive wealth in the hands of investors, which drives the price of some favored class of assets out of its normal relationship with the rest of the economy, and it ends in the economic sphere with the smoking crater left by the assets in question as their price plunges roughly as far below the mean as it rose above it, dragging the rest of the economy with it. It's the middle of the trajectory that passes through the particular form of crowd psychology that drives speculative bubbles, and since this is outside the economic sphere, neoclassical economics fails to deal with it.

This would be no problem if neoclassical economists by and large recognized these limitations. A great many of them do not, and the result is the type of intellectual myopia in which theory trumps reality. Since neoclassical theory claims that economic decisions are made by individuals acting freely and rationally to maximize the benefits accruing to them, many economists seem to have convinced themselves that any economic decision, no matter how harshly constrained by political power or wildly distorted by the mob psychology of a speculative bubble in full roar, must be a free and rational decision that will allow individuals to maximize their own benefits, and will thus benefit society as a whole.

As mentioned earlier, those who practice this sort of purblind thinking often find it very lucrative to do so. Economists who urged more free trade on the Third World at a time when "free trade" distorted by inequalities of power between nations was beggaring the Third World, like economists who urged people to buy houses at a time when houses were preposterously overpriced and facing an imminent price collapse, commonly prospered by giving such appallingly bad advice. Still, it seems unreasonable to claim that all economists are motivated by greed, when the potent force of a habit of thinking that fails to deal with the economic impact of non-economic forces also pushes them in the same direction.

That same pressure, with the same financial incentives to back it up, also drives the equally bad advice so many neoclassical economists are offering governments and businesses about the future of fossil fuels. The geological and thermodynamic limits to energy growth, like political power and the mob psychology of bubbles, lie outside the economic sphere. The interaction of economic processes with energy resources creates another end run: extraction of fossil fuels to run the world's economies, an economic process, drives the depletion of oil and other fossil fuel reserves, a non-economic process, and this has already proven its power to flow back into the economic sphere in the form of disastrous economic troubles. Once again, the inability to make sense of the interactions between economic activity and the rest of the world consistently blindsides contemporary economic thought.

Harnessing Hippogriffs

This same blindness to non-economic factors also affects another of the fundamental assumptions of modern economics, the law of supply and demand. According to this law, the supply of any commodity available in a free market is controlled by the demand for that commodity.

The law is supposed to work like this: When consumers want more of a commodity than is available on the market, and are willing to pay more for it, the price of the commodity goes up; this provides an economic incentive for producers to produce more of the commodity, and so the amount of the commodity on the market goes up. Increased production sets an upper limit on price increases, since producers competing against one another will cut prices to gain market share, and the willingness of consumers to pay rising prices is also limited. Thus, in theory, the production and price of a commodity are entirely determined by the balance between the desire of consumers to buy it and the desire of producers to make a profit from producing it.

This process is the "invisible hand" of Adam Smith's economic theory, the summing up of individual economic decisions to guide the market as a whole. Within certain limits, and in certain circumstances, the law of supply and demand works tolerably well. The problem creeps in when economists lose track of the existence of those limits and circumstances, and this, to a remarkable degree, is exactly what most contemporary economists have done.

As a result, even those branches of economic thinking that ought to take physical limits into account have come to treat money as a supernatural force that can conjure resources out of thin air. The most important example just now, as already suggested, is the way conventional economics treats energy. It's an article of faith

among the great majority of today's economists that the supply
of energy in the industrial world is purely a function of the law
of supply and demand. This article of faith has remained fixed in
place even as world energy supplies have plateaued in recent years,
and the most crucial of all energy supplies—the supply of petroleum, which provides some 40 percent of the world's energy and
effectively all its transportation fuel—peaked in 2005 and has been
slowly declining ever since.[7]

The resulting mismatch between theory and practice can approach the surreal. Consider the estimates of future petroleum
production circulated by the Energy Information Administration (EIA), a branch of the US government. Those estimates have
consistently predicted that petroleum production would go up
indefinitely. The logic behind these predictions, stated in so many
words in EIA publications, is the assumption that as demand for
petroleum goes up, supply will automatically keep pace with it.
The most recent estimates have kept the supply of petroleum in
step with rising demand, despite the decline in known sources, by
inserting a category labeled "unidentified projects"—predicting by
2030 no less than 43 million barrels a day of "unidentified projects,"
comprising around a half of total world production by that date.[8]
These "unidentified projects" are nowhere to be seen in the real
world; their sole purpose is to make reality fit the requirements of
economic theory, at least on paper. Energy blogger Kurt Cobb has
aptly labeled this sort of thinking "faith-based economics."[9]

The faith in question remains cemented in place across most of
contemporary economic thought. Whenever an economist enters
the debate about the future of world energy supplies, it's a safe bet
that he or she will claim that geological limits to the world's petroleum supply don't matter, because the invisible hand of the market
will inevitably solve any shortfall that happens to emerge. There's
a rich irony here, for shortfalls began to emerge promptly after
the world passed its peak of conventional petroleum production

in 2005; economists responded to those shortfalls by insisting that declining production is simply a sign that the demand for fossil fuel energy has decreased. No doubt when people are starving in the streets, we will hear claims that this is simply a reflection of the fact that the demand for food has dropped.

It may be worth exploring an extreme counterexample in order to clarify the limits to the law of supply and demand. Imagine that a plane full of economists makes a forced landing in the Pacific close to a desert island. The island has no food, no water and no shelter; it's just a bare lump of rock and sand with a few salt-tolerant grasses on it. As the economists struggle ashore from the sinking plane, the demand for food, water, and shelter on that island is going to be considerable, but even if each of the economists has a million dollars in his or her briefcase, that demand is going to go unfilled, until and unless a ship arrives with supplies from somewhere else. The lesson here is simple: economics does not trump physical reality.

More generally, the theoretical relationship between supply and demand functions only when supply is not constrained by factors outside the economic sphere. The constraints in question can be physical: no matter how much money you're willing to pay for a perpetual motion machine, for instance, you can't have one, because the laws of thermodynamics don't take bribes. They may be political: Nazi Germany had a large demand for oil from 1943 to 1945, for example, and the Allies had plenty of oil to sell, but anyone who assumed on that basis that a deal would be cut was in for a big disappointment. They may be technical: no matter how much you spend on providing health care for an individual, for instance, sooner or later it will be of no use, because nobody's yet been able to develop an effective cure for death.

Economists have come up with various workarounds to deal with external factors of this sort, some more convincing than others, but an inability to see economics as a subset of a much

larger world governed by non-economic forces remains endemic
to the discipline, and has caused some of its more spectacular failures. That inability undermines the theory of free markets governed by supply and demand: however pleasant free markets look
on paper, they do not exist. Strictly speaking, they are as mythical
as hippogriffs.

It occurs to me that some of my readers may not be as familiar
with hippogriffs as they ought to be. For those who lack so basic
an element of their education, a hippogriff is the offspring of a
gryphon and a mare; it has the head, body, hind legs and tail of a
horse, and the forelimbs and wings of a giant eagle. Hippogriffs are
said to be the strongest and swiftest of all flying creatures, which
is why Astolpho rode one to the terrestrial paradise to recover
Orlando's lost wits in Lodovico Ariosto's great poem *Orlando Fu-
rioso*. Their only disadvantage, really, is the minor point that they
don't happen to exist, and planning to use them as a new, energy-
efficient means of air transport, for instance, will inevitably come
to grief on that annoying little detail.

Free markets are subject to the same problem. There have been
many examples of market economies in history that were not controlled by governments, but there have been no examples of market
economies that were not controlled at all, and if one were to be set
up, it would remain a free market for maybe a week at most. Adam
Smith himself explained why, in memorable language: "People of
the same trade seldom meet together, even for merriment and di-
version, but the conversation ends in a conspiracy against the public, or some contrivance to raise prices."[10] When a market is not
controlled by government edicts, religious taboos, social customs
or some other outside force, it will quickly be controlled by com-
binations of individuals whose wealth and strategic position in the
market enable them to maximize the economic benefits accruing
to them by squeezing out rivals, manipulating prices, buying up
their suppliers, bribing government officials and the like—that is

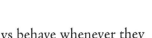

to say, behaving the way capitalists always behave whenever they are left to their own devices. This is what created the profoundly dysfunctional economy of Gilded Age America, and it also played a very large role in setting up today's economic troubles.

There's a rich irony here, in that the market economy portrayed in textbooks — in which buyers and sellers are numerous and independent enough that free competition regulates their interactions — is a form of commons, and a great many people who claim to be advocates of the free market have spent years arguing that commons should be eliminated wholesale in favor of private ownership. All commons systems, as Garrett Hardin pointed out in a famous essay,[11] have to be managed in ways that prevent individuals from exploiting the commons for their own private benefit; otherwise they fail. The 2009 Nobel laureate in economics, Elinor Ostrom, won her award for demonstrating that it's entirely possible to manage a commons so that Hardin's "tragedy of the commons" does not happen,[12] and she's quite right — there have been many examples of successfully managed commons in history. Strip away the management that keeps it from being abused, however, and the market, like any other commons, destroys itself.

The Market as Commons

Heretical as this concept may be in term of contemporary economics, it can be demonstrated easily enough by a piece of economic history within the experience of many of my readers. Consider an old-fashioned shoe store of the sort most American towns of any size had 50 years ago. Most of the profit that paid the store's bills and kept its proprietor fed and housed came from shoes of a relatively modest number of standard types and common sizes, but the store also carried types and sizes that were of special interest to local customers, including some quite unusual ones — for example, orthopedic shoes for a regular customer with foot deformities — and also provided shoe repair and alterations and a range of shoe-related products tailored specifically for the needs of the local market.

Such businesses could be quite successful, but the fact that most of their profit came from a fraction of their total line of goods and services made them vulnerable to competition from businesses that offered only those profitable lines. This was the opening that department stores exploited in the postwar years. As they expanded out of the large cities, they provided shoes in the more popular styles and sizes, without the other goods and services the local shoe stores provided. Since the department stores did not have the costs associated with these other goods and services, and could draw on economies of scale out of reach of local shoe stores, they could sell their products more cheaply than the local shoe stores.

As a result, customers went to the department stores rather than the local shoe stores and many of the latter were forced out of business. This meant that many of the specialized goods and services that had once been available in towns across America — in this example, shoe repair — stopped being available, except to those willing to travel to a city large enough for a shoe repair store to remain viable on its own, without the income stream the old stores had received from the sale of common varieties of shoes.

Nor does the process stop there, for the department stores turned out to be equally vulnerable to discount shoe stores, which provided even fewer services and an even more restricted range of styles and sizes, and thus either outcompeted the department store shoe departments or forced them to follow the discount store model. The result, in most North American towns and a surprisingly large number of cities, is that the only shoes available to consumers at all are cheaply made, poorly fitting mass-produced shoes in a small range of styles and sizes, sold in discount shops by clerks who wouldn't know how to help a customer find a shoe that fits even if that were part of their job description.

This logic is by no means limited to shoes. The same race to the bottom in which quality goods and services become unavailable and local communities suffer has taken place in nearly the same way in nearly every industry in the industrial world. A torrent of cheap shoddy goods funneled through Wal-Mart and its ilk have driven local businesses out of existence and made the superior products and services once provided by those businesses effectively unavailable to a great many people. In theory, this produces a business environment that is more efficient and innovative; in practice, the efficiencies are by no means clear and the innovation seems mostly to involve the creation of ever more exotic and unstable financial instruments — not necessarily the sort of thing that our society is better off encouraging.

In the real world, in other words, a laissez-faire market doesn't always produce improved access to better and cheaper goods and services, as Smith argued it would;[13] instead, it can put desirable products out of reach of consumers who would be happy to pay for them, but are not numerous enough to generate enough business to keep a shop from shutting its doors. It can also have disastrous impacts on such non-economic goods as healthy communities. The shift from an economy of local firms, which spent most of their income locally, to an economy of multinational firms that effectively pump money out of most of the world's communities to concentrate it in a handful of important cities, has played a massive role in the economic debacle that has overwhelmed so many towns and rural regions in the industrial world.

These effects can be understood by recognizing that a market is a commons, along the lines sketched out by the Garrett Hardin essay mentioned above. Like any other commons, it can break down when it is not managed in ways that keep the common interest of all participants from being harmed by the actions of individuals. This does not mean that markets ought to be abolished, any more than Hardin's arguments show that commons ought to be abolished; the idea that markets ought to be replaced by government bureaucracies was tested thoroughly in the Marxist states of the twentieth century and turned out to be a comprehensive flop. What it means, as I propose to show later on in this book, is that the same logic of checks and balances that has proven to work tolerably well in the political sphere needs to be applied to the economic sphere, particularly to those dimensions of economics that overlap with non-economic realities.

Energy's Rules

Despite the problems just outlined, the faith in free markets governed by supply and demand remains central not only to contemporary economics but to much of the modern world's collective conversation about the future. It's very common, for example, to hear well-intentioned people insist that the market, as a matter of course, will respond to restricted fossil fuel production by channeling investment funds either into more effective means of producing fossil fuels, on the one hand, or new energy sources on the other. The logic seems impeccable at first glance: as the price of oil, for example, goes up, the profit to be made by bringing more oil or oil substitutes onto the market goes up as well; investors eager to maximize their profits will therefore pour money into ventures producing oil and oil substitutes, and production will rise accordingly until the price comes back down.

That logic owes much of its influence to the fact that in many cases, markets do behave this way. Like any description of a complex system, though, the use of the invisible hand as an explanatory tool needs to be balanced by an awareness of the situations in which it fails to work. The economics of energy defines one of these situations. Energy, as E. F. Schumacher pointed out,[14] is not simply one commodity among others; it is the ur-commodity, the foundation for all economic activity. It follows laws of its own — the laws of thermodynamics — which are not the same as the laws of economics, and when the two sets of laws come into conflict, the laws of thermodynamics win every time.[15]

For a useful example, consider an agrarian civilization that runs on sunlight, as every human society did until the rise of industrialism three centuries ago. In energetic terms, part of the annual influx of solar energy is collected via agriculture, stored as grain and transformed into mechanical energy by feeding the grain to human laborers and draft animals. It's an efficient and resilient system, and under suitable conditions it can deploy astonishing amounts of energy; the Great Pyramid is one of the more obvious pieces of evidence for this fact.

Agrarian civilizations of this kind very often develop thriving market economies in which goods and services are exchanged between individuals. They also develop intricate systems of social abstractions that manage the distribution of these goods and services among their citizens. Both these, however, depend on the continued energy flow from sun to fields to granaries to human and animal labor forces. If something interrupts this flow — say, a failure of the annual grain harvest — the only option that allows for collective survival is to have enough solar energy stored in the granaries to take up the slack.

This is necessary because energy doesn't follow the ordinary rules of economic exchange. Most other commodities still exist after they've been exchanged for something else, and this makes exchanges reversible; for example, if you're the pharaoh of Egypt and you sell gold to buy marble for your latest pyramid, and then change your mind, you can normally turn around and sell marble to buy gold. The invisible hand works here: if marble is in short supply, those who have gold and want marble may have to offer more gold for their choice of building materials, but the marble quarries will soon be working overtime to balance things out.

Energy is different. Once you turn the energy content of a few million bushels of grain into a pyramid, say, by using the grain to feed workers who cut and haul the stones, that energy is gone, and you cannot turn the pyramid back into grain; all you can do is wait until the next harvest. If that harvest fails, and the stored energy in the granaries has already been turned into pyramids, neither the market economy of goods and services or the abstract system of distributing goods and services can make up for it. Nor, of course, can you send an extra ten thousand workers into the fields if you don't have the grain to keep them alive.

The peoples of agrarian civilizations generally understood this. It's part of the tragedy of the modern world that most people nowadays do not, even though our situation is not all that different from theirs. We're just as dependent on energy inputs from Nature, though our inputs include vast quantities of prehistoric sunlight, in the form of fossil fuels, as well as current solar energy in various forms. Atop that foundation, we have built our own kind of markets to exchange goods and services, and an abstract system for managing the distribution of goods and services — money — that is as heavily wrapped in mythology as anything created by the archaic agrarian civilizations of the past.

The particular form taken by money in the modern world has certain effects, however, not found in ancient systems. In the old agrarian civilizations, wealth consisted primarily of farmland and its products. The amount of farmland in a kingdom might increase slightly through conquest of neighboring territory or investment in canal systems, though it might equally decrease if a war went badly or canals got wrecked by sandstorms. Everybody hoped when the seed grain went into the fields that the result would be a bumper crop, but no one imagined that the grain stockpiled in the granaries would somehow multiply itself over time. Nowadays, by contrast, it's assumed as a matter of course that money ought automatically to produce more money.

That habit of thought has its roots in the three centuries of explosive economic growth that followed the birth of the industrial age. In an expanding economy, the amount of money in circulation needs to expand fast enough to roughly match the expansion in the range of goods and services for sale; when this fails to occur, the shortfall drives up interest rates (that is, the cost of using money) and can cause economic contraction. This was a serious and recurring problem across the industrial world in the nineteenth century, and led reformers in the Progressive era to reshape industrial economies in ways that permitted the money supply to expand over time to match the expectation of growth. Once again, the invisible hand was at work, with some help from legislators: a demand for an expanding money supply eventually gave rise to a system that built a constantly expanding money supply into the foundations of its economy.

That system, taken very nearly to its furthest possible extreme, is the economy that exists today in most nations of the industrial world. Created in response to an age of unparalleled growth, it assumes that perpetual growth on the same scale is an inevitable fact of economic life. The notion that growth might turn out to be a temporary, if protracted, phenomenon of the recent past, and will not continue into the future, will be found nowhere in contemporary mainstream economics or politics. It's true, of course, that three centuries of statistics support the idea of perpetual growth; it's not often remembered that those three centuries represent a tiny and very unusual fraction of humanity's trajectory on this planet, but there is another problem with those numbers. These days, a very large proportion of the numbers are faked.

Lies and Statistics

An economy is a system for exchanging goods and services, with all the irreducible variability that this involves. How many potatoes are equal in value to one haircut, for example, varies a good deal, because no two potatoes and no two haircuts are exactly the same, and no two people can be counted on to place quite the same value on either one. The science of economics, however, is mostly about numbers that measure, in abstract terms, the exchange of potatoes and haircuts (and, of course, everything else).

Economists rely implicitly on the claim that those numbers have some meaningful relationship with what's actually going on when potato farmers get their hair cut and hairdressers order potato salad for lunch. As with any abstraction, a lot gets lost in the process, and sometimes what gets left out proves to be important enough to render the abstraction hopelessly misleading. That risk is hardwired into any process of mathematical modeling, of course, but there are at least two factors that can make it much worse.

The first is that the numbers can be deliberately juggled to support some agenda that has nothing to do with accurate portrayal of the underlying reality. The second, subtler and even more misleading, is that the presuppositions underlying the model can shape the choice of what's measured in ways that suppress what's actually going on in the underlying reality. Combine these two and what you get might best be described as speculative fiction mislabeled as useful data — and the combination is exactly what has happened to economic statistics.

For decades now, to begin with, the US government, like that of most other nations, has tinkered with economic figures to make unemployment look lower, inflation milder and the country more prosperous. The tinkerings in question are perhaps the most enthusiastically bipartisan program in recent memory, encouraged by administrations and congress people from both sides of the aisle, and for good reason: life is easier for politicians of every stripe if they can claim to have made the economy work better. As Bertram Gross predicted back in the 1970s,[16] economic indicators have been turned into "economic vindicators" that subordinate information to public relations, and the massaging of economic figures Gross foresaw has turned into cosmetic surgery on a scale that would have made the late Michael Jackson gulp in disbelief.[17]

When choices are guided by numbers, and the numbers are all going the right way, it takes a degree of insight unusual in contemporary life to remember that the numbers may not reflect what is actually going on in the real world. You might think that this wouldn't be the case if the people making the decisions know that the numbers are being fiddled with to make them more politically palatable, as economic statistics in the United States and elsewhere generally are.

It's important, therefore, to remember that we've gone a long way past the simplistic tampering with data practiced in, say, the Lyndon Johnson administration. With characteristic Texan straightforwardness, Johnson didn't leave statistics to chance; he was well known in Washington politics for sending any unwelcome number back to the bureau that produced it, as many times as necessary, until he got a figure he liked.

Nowadays nothing so crude is involved. The president — any president, of any party, or for that matter of any nation — simply expresses a hope that next quarter's numbers will improve; the head of the bureau in question takes that instruction back to the office; it goes down the bureaucratic food chain, and some anonymous staffer figures out a plausible reason why the way of calculating the numbers should be changed; the new formula is approved by the bureau's tame academics, rubberstamped by the appropriate officials, and goes into effect in time to boost the next quarter's numbers. It's all very professional and aboveboard, and the only sign that anything untoward is involved is that for the last 30 years, every new formulation of official economic statistics has made the numbers look rosier than the one it replaced.

It's entirely possible, for that matter, that a good many of those changes took place without any overt pressure from the top at all. Hagbard's Law is a massive factor in modern societies. Coined by Robert Shea and Robert Anton Wilson in their tremendous satire *Illuminatus!*, Hagbard's Law states that communication is only possible between equals. In a hierarchy, those in inferior positions face very strong incentives to tell their superiors what the superiors want to hear rather than 'fessing up to the truth. The more levels of hierarchy between those who gather information and those who make decisions, the more communication tends to be blocked by Hagbard's Law. In today's governments and corporations, the disconnect between the reality visible on the ground and the numbers viewed from the corner offices is as often as not total.

Whether deliberate or generated by Hagbard's Law, the manipulation of economic data by the government has been duly pilloried in the blogosphere, as well as the handful of print media willing to tread on such unpopular ground. Still, I'm not at all sure these deliberate falsifications are as misleading as another set of distortions. When unemployment figures hold steady or sink modestly, but you and everyone you know are out of a job, it's at least obvious that something has gone haywire. Far more subtle, because less noticeable, are the biases that creep in because people are watching the wrong set of numbers entirely.

Consider the fuss made in economic circles about productivity. When productivity goes up, politicians and executives preen themselves; when it goes down, or even when it doesn't increase as fast as current theory says it should, the cry goes up for more government largesse to get it rising again. Everyone wants the economy to be more productive, right? The devil, though, has his usual residence among the details, because the statistic used to measure productivity doesn't actually measure how productive the economy is.

By productivity, economists mean *labor productivity*—that is, how much value is created per unit of labor. Thus anything that cuts the number of employee hours needed to produce a given quantity of goods and services counts as an increase in productivity, whether or not it is efficient or productive in any other sense.

Here's what *A Concise Guide to Macroeconomics* by Harvard Business School professor David A. Moss, as mainstream a book on economics as you'll find anywhere, has to say about it: "The word [productivity] is commonly used as a shorthand for labor productivity, defined as output per worker hour (or, in some cases, as output per worker)."[18]

Output, here as always, is measured in money — usually, though not always, corrected for inflation — so what "productivity" means in practice is income per worker hour. Are there ways for a business to cut down on the employee hours per unit of income without actually becoming more productive in any meaningful sense? Of course, and most of them have been aggressively pursued in the hope of parading the magic number of a productivity increase before stockholders and the public.

Driving the fixation on labor productivity is the simple fact that in the industrial world, for the last century or so, labor costs have been the single largest expense for most business enterprises, in large part because of the upward pressure on living standards caused by the impact of cheap abundant energy on the economy. The result is a close parallel to Liebig's law of the minimum, one of the core principles of ecology. Liebig's law holds that the nutrient in shortest supply puts a ceiling on the growth of living things, irrespective of the availability of anything more abundant. In the same way, our economic thinking has evolved to treat the costliest resource to hand, human labor, as the main limitation to economic growth, and to treat anything that decreases the amount of labor as an economic gain.

Yet if productivity is treated purely as a matter of income per worker hour, the simplest way to increase productivity is to change over from products that require high inputs of labor per dollar of value to those that require less. As a very rough generalization, manufacturing goods requires more labor input overall than providing services, and the biggest payoff per worker hour of all is in financial services — how much labor does it take, for example, to produce a credit swap with a face value of ten million dollars?

An economy that produces more credit swaps and fewer potatoes is in almost any real sense less productive, since the only value credit swaps have is that they can, under certain arbitrary conditions, be converted into funds that can buy concrete goods and services, such as potatoes. By the standards of productivity universal in the industrial world these days, however, replacing potato farmers with whatever you call the people who manufacture credit swaps counts as an increase in productivity. If you have been wondering why so many corporations with no obvious connection to the world of finance, such as auto manufacturers, launched large financial branches in recent decades, this is part of the reason why: the higher income per worker hour from manufacturing financial paper enables the firm to claim increases in productivity.

As important as the misinformation produced by these inappropriate statistical measurements, however, is the void that results because more important figures are not being collected at all. In an age that will increasingly be constrained by energy limits, for example, a more useful measure of productivity might be energy productivity — that is, output per barrel of oil equivalent (BOE) of energy consumed. An economy that produces more value with less energy input is an economy better suited to a future of energy constraints, and the relative position of different nations, to say nothing of the trend line of their energy productivity over time, would provide useful information to governments, investors and the general public alike.

Even when energy was still cheap and abundant, the fixation on labor productivity was awash with mordant irony, because only in times of relatively robust economic growth did workers who were rendered surplus by such "productivity gains" readily find jobs elsewhere. At least as often, they added to the rolls of the unemployed, or pushed others onto those rolls, fueling the growth of an impoverished underclass. With the end of the age of cheap energy, though, the fixation on labor efficiency promises to become a millstone around the neck of the world's industrial economies.

Economic Superstitions

After all, a world that has nearly seven billion people on it and a dwindling supply of fossil fuels can do without the assumption that putting people out of work and replacing them with machines powered by fossil fuels is the way to prosperity. This is one of the unlearned lessons of the global economy that is now coming to an end around us. While it was billed by friends and foes alike as the triumph of corporate capitalism, globalization can more usefully be understood as an attempt to prop up the illusion of economic growth by transferring the production of goods and services to economies that are, by the standards just mentioned, less efficient than those of the industrial world. Outside the industrial nations, labor proved to be enough cheaper than energy that the result was profitable, and allowed industrial nations to maintain their inflated standards of living for a few more years.

At the same time, the brief heyday of the global economy was only made possible by a glut of petroleum that made transportation costs negligible. That glut is ending as world oil production begins to decline, while the Third World nations that profited most by globalization cash in their newfound wealth for a larger share of the world's energy resources, putting further pressure on a balance of power that is already tipping against the United States and its allies. The implications for the lifestyles of most Americans will not be welcome.

To extract ourselves from the corner into which we have backed ourselves, however, requires coming to terms with the fact that a very large number of the previous choices all of us have made were founded on folly. If we lived in a world in which people always made rational decisions to maximize benefits to themselves, recognizing our past folly would be simply another rational decision, but in the real world things are not quite so simple. To understand why, it's necessary to talk a little about the role of superstition in human affairs.

In the area where I live, the Appalachian mountains of eastern North America, superstition is very much a living phenomenon. Many local gardeners, for example, choose times to plant seeds according to the signs and phases of the moon. This habit may reasonably be considered a superstition, but that word has a subtler meaning than most people remember these days. A superstition is literally something "standing over" (in Latin, *super stitio*) from a previous age; more precisely, it's an observance that has become detached from its meaning over time. A great many of today's superstitions thus descend from the religious observances of archaic faiths. When my wife's Welsh great-grandmother set a dish of milk outside the back door for luck, for example, she likely had no idea that her pagan ancestors did the same thing as an offering to the local tutelary spirits.

Yet there's often a remarkable substrate of ecological common sense interwoven with such rites. If your livelihood depends on the fields around your hut, for example, and rodents are among the major threats you face, a ritual that will attract cats and other small predators to the vicinity of your back door night after night is not exactly foolish. The Japanese country folk who consider foxes the messengers of Inari the rice god, and put out offerings of fried tofu to attract them, are mixing agricultural ecology with folk religion in exactly the same way; in Japan, foxes are one of the main predators that control the population of agricultural pests. The logic behind planting by the signs is a bit more complex, but it may not be irrelevant that the sequence of signs include all the tasks needed to keep a garden or a farm thriving, more or less equally spaced around the lunar month, and a gardener who works by the signs can count on getting the whole sequence of gardening chores done in an order and a timing that consistently works well.

There's a lot of this sort of thing in the world of superstition. Nearly all cultures that get any significant amount of their food from hunting, for example, use divination to decide where to hunt on any given day. According to game theory, the best strategy in any competition has to include a random element in order to keep the other side guessing. Most prey animals are quite clever enough to figure out a nonrandom pattern of hunting—there's a reason why deer across America head into suburbs and towns, where hunting isn't allowed, as soon as hunting season opens each year—so inserting a random factor into hunting strategy pays off in increased kills over time. As far as we know, humans are the only animals that make decisions with the aid of horoscopes, tarot cards, yarrow stalks and the like, and it's intriguing to think that this habit may have had a significant role in our evolutionary success.

Is this all there is to the practice of superstition? It's a good question, but one that's effectively impossible to answer. For all I know, the ancient civilizations that built vast piles of stone to the honor of their gods may have been entirely right to say that Marduk, Osiris, Kukulcan et al. were well pleased by having big temples erected in their honor, and reciprocated by granting peace and prosperity to their worshippers. It may just be a coincidence that directing the boisterous energy of young men into some channel more constructive than street gangs or civil war is a significant social problem in most civilizations, and giving teams of young men huge blocks of stone to haul around, in hot competition with other teams, consistently seems to do the trick. It may also be a coincidence that convincing the very rich to redistribute their wealth by employing huge numbers of laborers on vanity buildings provides a steady boost to even the simplest urban economy. Maybe this is how Kukulcan shows that he's well pleased.

Still, there's a wild card in the deck, because it's possible for even the most useful superstition to become a major source of problems when conditions change. When the classic lowland Mayan civilization overshot the carrying capacity of its fragile environment, for example, the Mayan elite responded to the rising spiral of crisis by building more and bigger temples. That had worked in the past, but it failed to work this time, because the situation was different; the problem had stopped being one of managing social stresses within Mayan society, and turned into one of managing the collapsing relationship between Mayan society and the natural systems that supported it. This turned what had been an adaptive strategy into a disastrously maladaptive one, as resources and labor that might have been put to use in the struggle to maintain a failing agricultural system went instead to a final spasm of massive construction projects. This time, Kukulcan was not pleased, and lowland Mayan civilization came apart in a rolling collapse that turned a proud civilization into crumbling ruins.

Rationalists might suggest that this is what happens to a civilization that tries to manage its economic affairs by means of superstition. That may be so, but the habit in question didn't die out with the ancient lowland Mayans; it's alive and well today, with a slight difference. The Mayans built huge pyramids of stone; we build even vaster pyramids of money.

It's all too accurate these days to describe contemporary economics as a superstition in the strict sense of the word. The patterns of dysfunction summarized in this chapter—the factors inherent to the profession of economics that make for bad decisions; the blindness to the impact of non-economic factors on economic processes; the belief in the infallibility of free markets in the face of contrary evidence; the reliance on "cooked" and irrelevant statistics—are all part of a way of thinking about economic life that worked tolerably well, from certain perspectives, during the age of economic expansion that was kick-started by the Industrial Revolution and reached its peak in the late twentieth century. Like the Mayan habit of building pyramids, though, the reasons why it worked were not the reasons its votaries thought it worked, and underlying changes in the energy basis of the world's industrial economies have made today's economic superstitions a severe liability in the future bearing down on us.

Undead Money

Like most complex intellectual superstitions—consider astrology in the Middle Ages and Renaissance—economics has a particularly strong following among the political classes. Like every other superstition, in turn, it has a solid core of pragmatic wisdom to it, but that core has been overlaid with a great deal of somewhat questionable logic which does not necessarily relate to the real cause and effect relationships that link the superstition to its benefits. My wife's Welsh ancestors believed that the bowl of milk on the back stoop pleased the fairies and that's why the rats stayed away from the kitchen garden; the economists of the twentieth century, along much the same lines, believed that expanding the money supply pleased—well, the prosperity fairies, or something not too dissimilar—and that's why depressions stayed away from the United States.

In both cases it's arguable that something very different was going on. The gargantuan economic boom that made America the world's largest economy had plenty of causes. The strong regulations imposed on the financial industry in the wake of the Great Depression made a significant contribution (a point that will be explored in more detail later on in this book); the accident of political geography that kept America's industrial hinterlands from becoming war zones, while most other industrial nations got the stuffing pounded out of them, also had more than a little to do with the matter; but another crucial point, one too often neglected in studies of twentieth-century history, was the simple fact that the United States at mid-century produced more petroleum than all the other countries on Earth put together. The oceans of black gold on which the US floated to victory in two world wars defined the economic reality of an epoch. As a result, most of what passed for economic policy in the last 60 years or so amounted to attempts to figure out how to make use of unparalleled abundance.

That's still what today's economists are trying to do, using the same superstitious habits they adopted during the zenith of the age of oil. The problem is that this is no longer what economists need to be doing. With the coming of peak oil — the peak of worldwide oil production and the beginning of its decline — the challenge facing today's industrial societies is not managing abundance, but managing the end of abundance. The age of cheap energy now ending was a dramatic anomaly in historical terms, though not quite unprecedented; every so often, but rarely, it happens that a human society finds itself free from natural limits to prosperity and expansion — for a time. That time always ends, and the society has to relearn the lessons of more normal and less genial times. This is what we need to do now.

This is exactly what today's economics is unprepared to do, however. Like the lowland Mayan elite at the beginning of their downfall, our political classes are trying to meet unfamiliar problems with overfamiliar solutions, and the results have not been good. Repeated attempts to overcome economic stagnation by expanding access to credit have driven a series of destructive bubbles and busts, and efforts to maintain an inflated standard of living in the face of a slowly contracting real economy have heaped up gargantuan debts. Nor have these measures produced the return to prosperity they were expected to yield, and at this point finger-pointing and frantic pedaling in place seem to have replaced any more constructive response to a situation that is becoming more dangerous by the day.

The sheer scale of the debt load on the world's economies is an important part of the problem. Right now, the current theoretical value of all the paper wealth in the world — counting everything from dollar bills in wallets to derivatives of derivatives of derivatives of fraudulent mortgage loans in bank vaults — is several orders of magnitude greater than the current value of all the actual goods and services in the world. Almost all of that paper wealth consists of debt in one form or another, and the mismatch between the scale of the debt and the much smaller scale of the global economy's assets means exactly the same thing that the same mismatch would mean to a household: imminent bankruptcy. That can take place in either of two ways — most of the debt will lose all its value by way of default, or all of the debt will lose most of its value by way of hyperinflation — or, more likely, by a ragged combination of the two, affecting different regions and economic sectors at different times.

What that implies for the not-too-distant future is that any economic activity that depends on money will face drastic uncertainties, instabilities and risks. People use money because it gives them a way to exchange their labor for goods and services, and because it allows them to store value in a relatively stable and secure form. Both these, in turn, depend on the assumption that a dollar has the same value as any other dollar, and will have roughly the same value tomorrow that it does today.

The mismatch between money and the rest of economic life throws all these assumptions into question. Right now there are a great many dollars in the global economy that are no longer worth the same as any other dollar. Consider the trillions of dollars' worth of essentially worthless real estate loans on the balance sheets of banks around the world. Governments allow banks to treat these as assets, but unless governments agree to take them, they can't be exchanged for anything else, because nobody in his right mind would buy them for more than a tiny fraction of their theoretical value. Those dollars have the same sort of weird half-existence that horror fiction assigns to zombies and vampires: they're undead money, lurking in the shadowy crypts of the world's large banks like so many brides of Dracula, because the broad daylight of the market would kill them at once.

It's been popular for some years, since the sheer amount of un-dead money stalking the midnight streets of the world's financial centers became impossible to ignore, to suggest that the entire sys-tem will come to a messy end soon in some fiscal equivalent of a zombie apocalypse movie. Still, the world's governments are doing everything in their not inconsiderable power to keep that from happening. Letting banks meet capital requirements with techni-cally worthless securities is only one of the maneuvers that govern-ment regulators around the world allow without blinking. Driving this spectacular lapse of fiscal probity, of course, is the awkward fact that governments — to say nothing of large majorities of the voters who elect them — have been propping up budgets for years with their own zombie hordes of undead money.

The only response to the current economic crisis most govern-ments can imagine involves churning out yet more undead money, in the form of an almost unimaginable torrent of debt; the only re-sponse most voters can imagine, in turn, involves finding yet more ways to spend more money than they happen to earn. So we're all in this together, guiding our actions by superstitions that no longer have any relation to the world in which we live. Everybody insists that the walking corpses in the basement are fine upstanding citi-zens, and we all pretend not to notice that more and more people are having their necks bitten or their brains devoured.

As long as most people continue to play along, it's entirely pos-sible that things could stumble along this way for quite a while, with stock market crashes, sovereign debt crises and corporate bankruptcies quickly covered up by further outpourings of unpay-able debt. The problem for individuals and families, though, is that all this makes money increasingly difficult to use as a medium of exchange or a store of wealth. If hyperinflation turns out to be the mode of fiscal implosion du jour, it becomes annoying to have to sprint to the grocery store with your paycheck before the price of milk rises above one million dollars a gallon; if we get deflationary contraction instead, business failures and plummeting wages make getting any paycheck at all increasingly challenging. In either case pensions, savings and insurance policies are as good as lost.

The act of faith that leads policy makers today to think that policies that failed last year will succeed next year is only part of the problem. The superstitions that lead so many intelligent people to think that our problems can be solved by pursuing new and expensive technological projects are another part. There are technologies that can help us right now, as I hope to show later on in this book, but they fall on the other end of the spectrum from the fusion reactors, solar satellites and plans to turn all of Nevada into one big algae farm that get so much attention today. Local, resilient, sustainable and cheap: these need to be our keywords for technological innovation just now. There are plenty of technological solutions that answer to that description, but again, our superstitions stand in the way.

In an age after abundance, the most deeply rooted of our superstitions — the belief that Nature can be ignored with impunity — is also the most dangerous. It's only fair to point out that for most people in the industrial world, for most of a century now, it has been possible to get away with this kind of thinking more often than not; the same exuberant abundance that produced ski slopes in Dubai and fresh strawberries in British supermarkets in January made it reasonable, for a while, to act as though whatever Nature tossed our way could be brushed aside. In an age after abundance, though, this may be the most dangerous superstition of all. The tide of cheap abundant energy that has defined our attitudes as much as our technologies is ebbing now, and we are rapidly losing the margin of error that made our former arrogance possible.

As that change unfolds, it might be worth suggesting that it's time to discard our current superstitions concerning economics, energy and Nature, and replace them with a more functional approach to these things. A superstition, once again, is an observance that has become detached from its meaning, and one of the more drastic ways this detachment can take place is through a change in the circumstances that make that meaning relevant. This has arguably happened to our economic convictions, and to a great many more of the commonplaces of modern thought; it's simply our bad luck, so to speak, that the consequences of pursuing those superstitions in the emerging world of scarcity and contraction are likely to be considerably more destructive than those of planting by the signs or leaving a dish of milk on the back step.

The remaining chapters of this book will attempt to sketch out some of the ways our current economic superstitions might best be replaced with more productive ways of understanding the production and exchange of goods and services among human beings. To make any progress toward that goal, however, it's necessary to realize that the production and exchange of goods and services among human beings is a subset, and a fairly small one, of a much larger economy that embraces the entire natural world. To grasp that, it's necessary to take the challenge to conventional economic thought a good deal deeper than we have taken it so far.

THE THREE ECONOMIES

ONCE AGAIN, we can begin this exploration with Adam Smith. *The Wealth of Nations* begins with the following sentence: "The annual labor of every nation is the fund which originally supplies it with all the necessities and conveniences of life." This same concept, variously phrased, forms one of the least questioned assumptions in modern economics; even most of those who dispute it offer what are at most slight variations — arguing, for example, that the labor of previous years embodied in capital is also crucial to understanding the economic process. Left unrecognized is the crucial fact that the annual labor of a nation would be utterly useless without the goods and services provided free of charge by Nature, which enable labor to be done at all by making human life possible in the first place and by providing all that labor with something to labor on.

This recognition has not simply been missed by economists; as often as not, it has been flatly rejected. One classic example is David Ricardo (1772–1823), one of Adam Smith's most influential successors. Ricardo remains a popular figure in economics, not least because his arguments on behalf of free trade arrangements proved to be highly useful to the British Empire in its time, and of course to the American Empire in ours; you can still find his arguments on this subject presented as simple fact in the pages of most freshman economics textbooks.[1] Another of the core elements of Ricardo's economic theories, though, is the claim that land retains its "original and indestructible" economic value no matter what economic use is made of it.[2]

This is an odd claim. Even in the early nineteenth century, when Ricardo originally made it, plenty of people could have set him straight; the fact that bad farming practices could make soil useless for farming was well known in Ricardo's time, and so was the impact of industrial pollution—though of course we have gained a great deal more bitter experience with both since then. Whatever the reasons for his claim, Ricardo's ideas concerning land prefigured the way that natural resources have been treated by most economists ever since. This is as true of radical economists as of their capitalist rivals; recent proponents of "green socialism," for example, might find it useful to reread Marx, who explicitly rejected the idea that the "free gifts of Nature" could have any value at all.[3] (The disastrous mistreatment of the environment common under Marxist regimes in the twentieth century was thus not accidental, but a logically necessary outgrowth of Marxist theory.) Nearly the only concession made to the ecological dimensions of economics in the mainstream, and it's a fairly recent one, is the concept of "externalities"—the recognition that if somebody does something that fouls the environment, other people may suffer a loss of economic value as a result, and might deserve compensation for that.[4]

Now of course this is true, and Garrett Hardin's theory of the tragedy of the commons built on that insight to remind us that a society that permits the advantages of ecological abuse to go to individuals, while the costs are shared by the whole society, is effectively subsidizing the destruction of its environment. Still, both the "externalities" argument and the structure Hardin built on it miss the central issues raised by the interface between the environment and economics. Both tacitly accept Ricardo's fantasy of invulnerable land as the normal state of affairs, apply it to the entire environment, and then focus attention on those supposedly exceptional situations when somebody does manage to make land (or some other environmental resource) less valuable.

To show where this thinking falls short, let's take a closer look at the land whose value Ricardo considered "indestructible." He was talking primarily about land as an economic factor in agriculture, and so shall we. What he apparently did not realize, but every country farmer knew in his time — and ecologists have demonstrated in fine detail in ours — is that fertile land suitable for growing crops does not simply happen. Like anything else of value, it must be made, and once made, it must be maintained. The only thing that sets it apart from the products of human industry is that the vast majority of the labor needed to make and maintain agricultural land is not performed by human beings.

Soil suitable for crops, after all, is not simply rock dust; pound for pound, it is among the most complex substances in the known universe. A large part of it — in the best soil, well over half — is organic matter, some living, some dead but not yet wholly decayed, some dissolved into organic colloids complex enough to give analytic chemists sleepless nights. All of these are put there by the activity of living beings over long periods of time. Energy and raw materials flow constantly through soil, uniting bacteria, fungi, algae, worms, insects and many other living beings into one of the most intricate ecosystems on Earth. Plants participate in and depend on this bewilderingly complex world; they draw water and mineral nutrients from it, and cycle leaves, root fibers and a wide range of chemical compounds back into it.[5]

The farmer who wants to grow crops is attempting to extract wealth from the underground ecosystem of the soil. She can ignore that, and simply plant and harvest with no attention to the needs of the soil, but if she does, the soil will be depleted of nutrients in a few years and her crops will fail. Alternatively, she can replace nutrients with chemical fertilizers, predators with pesticides, and so on. If she does this she will have to use steadily larger doses of chemicals to get the same yields, and if the society she lives in runs short on petroleum and natural gas feedstocks for these chemicals — as ours shows every sign of doing — she will be left with soil too sterile and pest ridden to grow anything. If she wants to fulfill Ricardo's promise and hand the land on to her grandchildren in the same condition that she received it from her grandparents, she will have to provide the things the soil needs for its long-term health. Put another way, she will have to barter with the soil, giving it the things it will accept in exchange for crops.

This is the premise of organic agriculture, of course. It's a premise that has proven itself in the Asian farming regions that inspired the organic pioneers of the early twentieth century to devise a more general system of agriculture that works with rather than against natural cycles,[6] and in the farms now using organic methods to get yields roughly comparable to those of chemical agriculture. The organic approach has many dimensions, but one may not have received the importance it deserves. To an organic farmer, land is not a commodity that can be owned but a community with which she interacts, *and that community has its own economy on which the farmer's economy depends.*

Imagine, to develop this concept into a metaphor, that our farmer got her crops, not from her fields, but from the village of an indigenous tribe near her home. The inhabitants of the village are deeply conservative, and their own economy follows traditional patterns not subject to change or negotiation. If the farmer wants crops, she must find out what the villagers are willing to take in exchange for them, and that will be determined by the internal dynamics of the village economy: what is already produced in surplus amounts, what is scarce, what is desired and what is detested by the villagers. Her relations with the village, in other words, would be exactly the same in outline as those of an organic farmer with her land.

The same thing is true of every other form of economic activity, though the dependence on Nature may be less obvious in some cases than in others. Behind the human activities that produce secondary goods lie a bewildering range of nonhuman activities — the biological cycles that yield soil fertility, crop pollination and countless other things of economic value; the hydrological cycles that put fresh water into reservoirs and taps; the tectonic processes in the crust that put economically useful metals and minerals into veins in the rocks; and, of central importance just now, the extraordinarily complex interplay of biological and geological processes that spent half a billion years storing away countless billions of tons of carbon under the earth's surface in the form of fossil fuels.

Conventional economics assumes that these things get there by some materialist equivalent of divine fiat. This disastrously misstates the situation. These natural goods are produced by an exact analogue of the way that secondary goods are produced: raw materials are transformed, through labor, using existing capital and available energy, to produce goods and services of value. The difference is that all this economic activity takes place in the nonhuman world. Human beings do not manage the production of natural goods, and the disastrous results of attempts to do so to date suggest that we probably never will. In at least some cases, however — maltreated farmland is a good example — we can interfere with the production of natural goods, and suffer the consequences when this mismanagement impacts our own economies.

What must be understood here is that human economic activity is far less independent of the natural world than too many economists try to pretend. The scale of this dependence is as rarely recognized as it is hard to overstate. One of the few attempts to quantify it, an attempt to work out the replacement costs for natural services carried out a few years back by a team headed by heretical economist Robert Costanza, came up with a midrange figure equal to around three times the gross domestic product of all human economic activity on earth.[7]

In other words, out of every dollar of value circulating in the world's human economies something like 75 cents were provided by natural processes rather than human labor. What's more, most if not all of that 75 cents of value had to be there in advance for the production of the other 25 cents to be possible at all. Before you can begin farming, for example, you need to have arable soil, water and an adequate growing season, as well as more specialized natural services such as pollination. These are nonnegotiable requirements; if you don't have them, you can't farm. The same is true of every other kind of productive work in the human economy: Nature's contribution comes first, and generally determines how much the human economy can produce.

The Power of Paradigms

Unfortunately, these reflections unfold from a way of thinking about the Nature of economic activity that is not shared by most people in the industrial world. The core of that way of thinking, and the focal point of the disagreements that surround it, is the issue of environmental limits. It's no exaggeration to say that either you believe in these limits or you don't. If you do, it seems glaringly obvious that modern industrial civilization, which depends on ever-increasing exploitation of finite and nonrenewable resources, is in deep trouble, and the only viable options are those that jettison the fantasy of perpetual growth and aim at a controlled descent to a level of energy and resource use per capita that can be sustained over the long run.

If you don't believe in limits, by contrast, such notions are the height of folly. Since, according to this way of thinking, progress can always overcome any limit Nature might impose on human beings, it seems glaringly obvious that modern industrial civilization needs to push progress into overdrive so that it can find and deploy the innovations that will get us past today's problems and launch our species onward toward its glorious future, whatever that happens to be.

Disbelief in environmental limits, as it happens, is far more common these days than belief in them. That's a fascinating twist of fate, because the evidence for the power of environmental limits over human life is overwhelming. Ecologists have documented the myriad ways that environmental limits play a dominant role in shaping the destiny of every species, ours included. Historians have chronicled the fate of many civilizations that believed themselves to be destiny's darlings, and proceeded to pave the road to collapse with their own ecological mistakes.[8] From a perspective informed by these facts, the insistence that limits don't apply to humanity is as good a case study as one might wish of that useful Greek word *hubris*, defined as the overweening pride of the doomed. Still, this makes it all the more intriguing that the power and relevance of environmental limits are anything but self-evident to most people in the industrial world today.

The power of nonrational assumptions in shaping human thought was mapped out decades ago by Thomas Kuhn, whose book *The Structure of Scientific Revolutions* is as famous as it is rarely read. Kuhn was among the first historians of science to put the popular image of scientific progress to the test, and he found it wanting. In place of the notion that science advances toward objective truth by the accumulation of proven facts — a notion that still shapes histories of science written for popular consumption — he showed that scientific beliefs are profoundly shaped by social and cultural forces, and that the relation between scientific theory and the facts on the ground is a great deal more complex than conventional ideas allow.

Kuhn's take on things has been misstated often enough that it probably needs a summary here. During a period of what he calls "normal science," scientists model their work on a paradigm. This isn't some sort of vague worldview, in the sense too often given to the word; rather, it's a specific example of science at work, an investigation in a given field by an exemplary scientist and the successful theory resulting from that research. In bacteriology, for example, Louis Pasteur's research program in the 1870s and 1880s, which led to the first artificial vaccines, became the paradigm that later researchers followed; good bacteriological research—in Kuhn's terms, normal science—was research that followed Pasteur's lead, fine-tuned his theories and asked the same kinds of questions about the same kinds of phenomena that he did.

Sooner or later, though, a mismatch opens up between the paradigm and the facts on the ground: the research methods drawn from the paradigm stop yielding good answers, and the paradigmatic theory no longer allows for successful prediction of phenomena. Scientists normally respond by pursuing the research methods with redoubled energy while making the theory more elaborate, the way that Ptolemy's earth-centered cosmology was padded out with epicycle after epicycle to make it fit the vagaries of planetary motion. Crisis comes when the theory becomes so cumbersome that even its stoutest believers come to realize that something is irreducibly wrong, or when data emerges that no reworking of the paradigmatic theory can explain. The crisis resolves when a researcher propounds a new theory that makes sense of the confusion. That theory, and the research program that created it, then becomes the new paradigm in the field.

So far, so good. Kuhn pointed out, though, that while the new paradigm solves questions the old one could not, the reverse is often true as well: the old paradigm does things the new paradigm cannot. It's standard practice for the new paradigm to include the value judgment that the questions the new paradigm answers are the ones that matter, and the questions the old paradigm answered better no longer count. Nor is this judgment pure propaganda; since the questions the new paradigm answers are generally the ones that researchers have been wrestling with for decades or centuries, they look more important than details that have been comfortably settled since time out of mind. They may also be more important, in every meaningful sense, if they allow practical problems to be solved that the old paradigm left insoluble.

Yet the result of that value judgment, Kuhn argued, is the false impression that science progresses by replacing false beliefs with more true ones, and thus gradually advances on the truth. He argued that different paradigms are not attempts to answer the same questions, differing in their level of accuracy, but attempts to answer entirely different questions — or, to put it another way, they are models that highlight different features of a complex reality, and cannot be reduced to one another. Thus, for example, Ptolemaic astronomy isn't wrong, just useful for different purposes than Copernican astronomy; if you want to know how the movements of the planets appear when seen from Earth — for the purposes of navigating a boat by the stars, for example — the Ptolemaic approach is still a better way to go about things.

These same considerations sprawl outside the limits of the sciences to define the rise and fall of paradigms in the entire range of human social phenomena. The difference between the believers and the disbelievers in limits is a difference in paradigms. Those who believe that modern industrial society is destined for, or capable of, unlimited economic expansion have drawn their paradigm from the Industrial Revolution and its three-century aftermath, with James Watt and his steam engine playing roughly the same role that Louis Pasteur played in the old paradigm of bacteriology. Like any other paradigm, the Industrial Revolution defines certain questions and issues as important and dismisses others from seri-ous consideration.

This is where the problems arise, because a solid case can be made that some of the questions dismissed from consideration by the "normal culture" of industrial expansion are those our species most needs to face just now, as the depletion of fossil fuel reserves and the soaring costs of environmental damage become central facts of our contemporary experience. The industrial paradigm can only interpret running out of one resource as a call to begin exploiting some even richer one. If there is no richer one, and even the poorer ones are rapidly being depleted as well, what then? From within the industrial paradigm, that question cannot even be formulated; the assumption that there is always some new and better resource to be had is hardwired into it.

Thus the current predicament of industrial society demands a change of paradigms. The belief in limits just discussed derives from a different model — the model of ecology, which is still sort-ing out its historical vision and has not yet quite found its paradig-matic theory, researcher and discovery. From within the emerging paradigm of ecology, the models that provide the most insight into our contemporary situation are those found in nonhuman Na-ture — above all, the cycles of increase, overshoot and dieoff which afflict so many other species that rely on outside forces to con-trol their numbers. The ecological paradigm suggests that unless we take that model and its implications into account, some of the most important factors shaping our future are completely out of sight.

The change from one paradigm to another, however, is not an overnight thing. Kuhn points out that in the sciences it usually has to wait until most of the older generation of scientists, who have been trained in the old paradigm, have been removed from the de-bate by old age and death. The same thing is too often true in other fields. Thus it's uncomfortably likely that even as the industrial paradigm fails to explain an increasingly challenging world, a great many people will cling to the faith that progress will bail us out, and ignore the fact that all the complex economic activities of the industrial world depend ultimately on Nature itself.

Primary and Secondary Goods

This is one of the many places where E. F. Schumacher's work provides a vital analytical tool. As mentioned in the Introduction, Schumacher made a distinction between what he called primary goods and secondary goods one of the foundations of his economic thought.[9] Secondary goods are the goods and services provided by human labor, the ordinary subject of economics as the discipline is currently practiced. Primary goods are the goods and services provided by Nature, and they make the production of secondary goods possible.

A failure to distinguish between primary and secondary goods is at the root of a great deal of today's economic nonsense. It's usually possible, for example, to substitute one secondary good for another if the supply of the second good runs short or the price gets too high, and for this reason it's a standard assumption of economics — and one of the foundations of the law of supply and demand — that consumers can meet their needs equally well with many different goods. It is rarely recognized that this assumption does not apply to natural, or primary, goods. In the world of Nature, a different rule — Liebig's law of the minimum, discussed in Chapter One — applies instead: production is limited by the scarcest necessary resource. Thus if you have a farm and can't get water for your crops, it doesn't matter if you have excellent soil and all the other requisites of farming; you can't grow anything.

In certain limited situations, to be sure, it's possible to substitute one primary good for another — for instance, to use low-grade iron ores such as taconite when high-grade ores have been exhausted by overenthusiastic mining. Even when this can be done, though, a law of diminishing returns always applies. You can get iron out of low-grade ore, but the extraction process is less efficient and takes much larger inputs of energy. When energy is cheap, you can ignore this — this is exactly what happened in the twentieth century, as the iron industry retooled itself to use steadily lower grades of ore and steadily larger inputs of energy — but that in itself simply passes costs onto the future, since the fossil fuels that provided the energy are themselves subject to depletion and to the law of diminishing returns. One way or another, the substitution imposes additional costs without providing any additional economic benefit.

This same rule also applies to every other natural good. Consider the valuable service provided to the world's economies by the honeybees that pollinate most nongrain food crops. If we succeed in adding the honeybee to the already long list of the world's recently extinct life forms, it would doubtless be possible to replace their pollination services by other means, whether that took the form of huge pollinating machines rumbling across the fields or the simpler and probably more economical approach of migrant workers using little brushes to wipe pollen from a bag onto the stamen of every single flower. Note, though, that while honeybees still exist, no farmer in his or her right mind would hire a thousand laborers with brushes instead of calling up the local beekeeper and arranging for a few hives to be left in the fields; substituting some other pollination method for bees would add a huge additional cost to farming, without yielding any additional benefit.

The failure to recognize the difference between secondary goods, which can be readily replaced by other goods without additional cost, and primary goods, which cannot, is among the most important forces driving our current economic troubles. For the last three centuries, the industrial economies of the world have been using up every primary good that can be converted into secondary goods, and doing so at extravagant and steadily increasing rates. Think of any good or service provided by Nature — from topsoil to oceanic fish stocks, from the pollution-absorbing capacities of rivers to the storm-buffering properties of wetlands, from breathable air and drinkable water to the mineral stocks and fossil fuel reserves that keep the entire system running—and you've just identified something that's being used up at a breakneck pace by industrial societies, with no thought of the potential costs of substituting something else for it, much less of the hard fact that nothing humanity can possibly do can provide a substitute for many of them once they're gone. As fossil fuel depletion adds a new round of substitution costs to those already in play, this same process will have even more dramatic impacts on the future.

Schumacher's insight that goods produced by Nature are the primary goods in any economy, and those produced by human labor are secondary goods, can thus usefully be extended further. There is also a primary and secondary economy. The cycles of Nature that produce goods needed by human beings constitute the primary economy, while the process by which human beings produce goods is the secondary economy. The recognition that Nature is an economy, not simply a source of raw materials and a dumping ground for "externalities," makes it easy to understand why replacing a depleted natural resource with something else always involves substitution costs: human labor must be brought in, and paid for, to replace some part of the work previously done by Nature, and the costs of that part of the work once done for free have to be paid out of the finite resources of the secondary economy.

We have become so used to thinking of economics as a matter of human labor that it's probably necessary to point out here that what are sometimes called "primary industries"—farming, mining and the like—belong to the secondary economy, not the primary one. The primary economy consists wholly of those nonhuman processes that yield economic goods to human beings. Thus a farm and the crops grown on it are part of the secondary economy, while the soil, water, sun and genetic potential in the seed stock that make the farm and its crops possible are part of the primary economy. In the same way, a mine belongs to the secondary economy, while the slow geological processes that put ore in the ground where it can be mined belong to the primary economy. Examine any human economic activity, and you'll find that it depends on natural processes that make that activity possible; those processes are the inputs from the primary economy that support the activities of the secondary economy.

Thus the famous dictum of Adam Smith cited earlier stands in desperate need of reformulation. The annual product of the natural environment of every nation is the fund which originally supplies it with all the necessities and conveniences of life; the annual human labor is simply the energy input required to turn some of that product into forms useful for human beings. The wealth of nations, it turns out, is ultimately the wealth of Nature.

Exchange Value and Nature's Wealth

At least an equal difference can be traced between primary and secondary goods, taken together, and a third class of goods which are produced neither by Nature nor by labor. These are tertiary or, more descriptively, financial goods; they form far and away the largest single class of goods in the world today in terms of monetary value, and the markets in which they are bought and sold dominate the economies of the world's industrial nations.

A specific example of a tertiary good may help make sense of this concept. Consider a corporate bond with a face value of $1,000. This is a good in the economic sense — there are people who want to buy it and people who are able to produce and sell it, and the buyers and sellers come together in a series of market exchanges in which such bonds pass from seller to buyer. Compare it to any more tangible item of value, though, and the bond is clearly a very strange sort of good. It consists of nothing more than a promise, on the part of the issuing corporation, to pay $1,000 at some future date. That promise may or may not be honored—junk bonds are bought and sold in full knowledge of the fact that the ability of the issuers to pay up is in serious doubt—but even then the chance of collecting on the bond is treated as an object of value.

The particular kind of value the bond has is called "exchange value," and it forms the keystone of contemporary economics. It's not an exaggeration to say that to today's economists, if the value of something can't be measured according to some form of exchange value, it has no value at all. In practice, since the basic yardstick of exchange value in modern economies is money, this amounts to the claim that if something can't be assigned a monetary value, it has no economic value at all. The widely held belief that the wealth of Nature is valueless until it's transformed into something else by human labor bases itself largely on the fact that nobody has to pay the nonhuman world for that wealth, and so figuring out the economic value of the products of natural systems poses a major challenge—not insoluble, but significant enough that few economists have been willing to take it up.

Now it's fair to note that the equation between value and exchange value, or more broadly between wealth and money, has formed a fault line running straight through the heartland of economic thought since Adam Smith's time. Some economists— these days, the great majority—treat wealth and exchange value measured in money as interchangeable concepts. Others—the minority nowadays—draw a sharp distinction between them. Those who accept the identity of money and wealth usually think of the rules governing money as something akin to laws of Nature, untainted by human purposes and agendas; those who draw a distinction between them see those rules as social constructs that benefit some people at the expense of others.

Over the span of human history, however, money is a fairly late invention, and until very recently it played only a small part in the lives of most people even in the societies that used it. Until the eighteenth century, even in the Western world, a majority of all goods and services were produced and exchanged within the household economy, or in traditional economies in which exchanges were governed by custom rather than market forces, and only the well off could expect to handle money on a daily basis. Furthermore, money functioned in significantly different ways when it consisted largely of marketable commodities, such as silver and gold, than it does in an economy that allows it to be created out of thin air by a few keystrokes.

It is true, of course, that every human society has had some social mechanism for distributing goods and services among its members. Paleoanthropologists have argued that it was precisely the evolution of food sharing within bands of ancestral humans that gave our species the evolutionary edge to expand across the globe in the face of wide variations in habitat and the rigors of ice age climates. Hunting and gathering societies around the world have intricate arrangements for sorting out who gets how much of the various natural sources of wealth available to them; so do the horticultural and pastoral human ecologies that evolved out of the hunter-gatherer pattern. Some of these latter use a particular trade good in certain contexts as a general marker for value — think of the shell-bead wampum strands used by the First Nations in eastern North America, for example — but these get used only in a restricted class of prestige exchanges, and play no role in everyday exchanges of goods and services. The same was true of gold and silver coinage in many ancient and medieval societies; most of the population of medieval England, for example, could expect to go from one winter to the next without seeing more than a handful of silver coins.

From a historical perspective, then, a money system of the sort we use today is simply one culturally specific way of managing the distribution of goods and services within a particular kind of human society. What sets money apart from other systems is not its convenience — quite the contrary; such alternatives as household production of goods and services or traditional economies of gift and customary exchange, are quite a bit more convenient for most purposes, since the extra steps imposed by the need to bring money into the situation can be done without. Rather, money has three distinctive features relevant to the present discussion.

First, it tends to draw all economic activity into its own ambit by supplanting other forms of exchange with monetary exchange. That can (and very often is) used for political control, but this is a side effect. The principal effect of this property of money is to turn a society into an economic monoculture.

Diversity is the basis of stability in any ecosystem, human or otherwise; when a significant proportion of goods and services are produced in the household economy without money changing hands, for example, the vagaries of the market economy have a limited influence on everyday life. That limitation goes away once goods formerly made at home have to be purchased in the market with money. Thus it's no accident that over the last four centuries, as the market has supplanted the household economy and other patterns of production and exchange, economic crises have become more frequent, more severe, and more widely felt. The effects of the Dutch tulip mania and the South Sea bubble were restricted to a relatively small proportion of their respective societies; this was hardly true of the Great Depression of the 1930s, and seems to be turning out even less true of the Great Recession now under way.

The second distinctive feature of a money economy is that it makes it harder, not easier, to value certain very large classes of goods. What Schumacher called primary goods are the most obvious example. Working out the value in money of primary goods is a real challenge in any money system; most traditional societies around the world, by contrast, have no trouble whatsoever recognizing the value of primary goods and finding ways to integrate that value into their own systems of exchange.

The salmon ceremonies of First Nations along the northwest coast of North America are a case in point. These societies have traditionally had a gift economy in which rank and social influence are gained by giving away goods — a system that once provided a very efficient means of circulating material wealth through their societies — and they treat the arrival of the annual salmon runs in exactly the same terms, as a mighty gift from the Salmon People that must receive an appropriate economic response.

Anthropologists who treat these arrangements purely as a matter of religion are missing a central aspect of their importance. They are, among other things, ways of integrating the traditional secondary economy into the primary economy of Nature, so that the value of the salmon harvest is always weighed in decisions that might affect it, and traditional practices that preserve salmon runs get potent economic sanction. Such arrangements are common — indeed, very nearly universal — in moneyless economies, and they can still be found in many historical economies that use money in a small way. The more completely an economy becomes subject to money, the more difficult it becomes to include primary goods in economic calculations. The Salmon People are perfectly capable of participating in a gift economy, but there's no way they can cash a check — or, for that matter, write one.

The third distinctive feature of money is subtler than these first two, and often misunderstood. Unlike other systems of distributing goods and services, money functions as a good in its own right, and the right to use it functions as a service. To some extent this is a legacy of the time when money was made of some culturally valued substance — wampum strings in eastern Native North America, say, or gold and silver in medieval Europe — but it opens the door to unexpected developments.

If money is treated as a good in its own right, and the use of money is treated as a service in its own right, then instead of exchanging money for ordinary goods and services and ordinary goods and services for money, it becomes possible and potentially profitable to exchange money for money. The entire world of finance, from savings accounts and installment loans up through the dizzying abstractions of today's derivative markets, unfolds from this third property of money. When money plays a relatively minor role in a society, this dimension is correspondingly small; as the volume and pervasiveness of money expands, so does the scale and impact of the arrangements by which money makes money. When money dominates a society, so does the world of finance, and the amount of money being traded for money can exceed by many orders of magnitude the amount of money being traded for goods and services.

What makes this problematic is that the rules governing money are not the same as those governing other goods and services. Unlike goods and services that have their own concrete value, money is only worth what it can buy; unlike goods and services that must be produced by labor from resources, money can be conjured from thin air by dozens of different kinds of financial alchemy, or by the momentary whim of a government. Nor does the amount of money in circulation have to have anything at all to do with the amount of other goods and services available. All these differences mean that the economy of money can very easily slip out of balance with the economy of nonfinancial goods and services.

It's useful here to extend the concept of tertiary wealth presented earlier in this chapter and speak of the economy of money as the *tertiary economy* of the modern world. If the primary economy consists of the natural processes that provide goods and services to human beings without human labor, and the secondary economy consists of the conjunction of human labor and natural goods that produces the goods and services Nature itself doesn't provide, the tertiary economy consists of the circulation of monetary goods and financial services that, in theory, fosters the distribution of the products of the primary and secondary economies, and in practice — at least at present — obscures crucial trends in the primary and secondary economies behind a fog of paper wealth.

The point that has to be grasped, in this as in so many other contexts, is that the three economies, and the three kinds of wealth they produce, are not interchangeable. Trillions of dollars in credit swaps and derivatives will not keep people from starving in the streets if there's no food being grown and no housing being built, or maintained or offered for sale or rent. The primary economy is fundamental to survival; the secondary economy is the source of all real wealth that doesn't come directly from Nature; the tertiary economy is simply a way of measuring wealth and managing its distribution; and treating these three very different things as though they are one and the same makes rank economic folly almost impossible to avoid.

The Anti-Ecology of Money

The differences between the tertiary economy and the primary and secondary economies run very deep, and those differences have consequences that are central to our current predicament. In the real world, the supply of tangible goods produced by natural cycles or human labor is limited by factors that may not necessarily respond to changes in demand. If there's only so much water in a river, for example, that's how much water there is; the fact that people want more, if such is the case, does not produce any more water than the hydrologic cycle is already willing to provide. Equally, if a country's labor force, capital plant and resource base are fully engaged in making a certain quantity of secondary goods, producing more requires a good deal more than a decision to do so; the country must increase its labor pool, its capital plant, its access to resources or some combination of these, in order to increase the supply of goods.

Yet the only limit on the production of tertiary goods is the demand for them. How many bonds can a corporation print?

For all practical purposes, as many as people are willing to buy. A good number of the colorful bankruptcies that have enlivened the business pages in recent years, for example, happened to firms that mistook a temporary bubble for permanent prosperity, issued bonds far beyond their ability to pay, and crashed and burned when all that debt started to come due. On an even more gargantuan scale, as I write these words, the United States government is trying to jumpstart its economy by spending money it doesn't have, and selling bonds to cover the difference. As a result, it is amassing debt on a scale that makes the most extravagant Third World kleptocracies look like pikers. It's hard to imagine any way in which the results of this absurd extravagance will be anything but ugly, and yet buyers around the world are still snapping up US treasury bonds as though there's a scintilla of hope they will see their money again.

The difference between the supply-limited goods of the primary and secondary economy and the demand-limited goods of the tertiary economy is, among other things, a difference between kinds of feedback. Think about a thermostat and it's easy to understand the principle at work here. When the temperature in the house goes below a certain threshold, the thermostat turns the heat on and brings the temperature back up; when the temperature goes above a higher threshold, the thermostat shuts the heat off and the temperature goes back down.

This is called negative feedback, and in a market economy all secondary goods are subject to negative feedback. This is the secret of Adam Smith's invisible hand: since the supply of any secondary good is limited by the available natural inputs, labor pool and capital stock, increased demand pushes up the price of the good, forcing some potential buyers out of the market, while decreased demand causes the good to become less expensive and allows more buyers back into the market. Equally, rising prices encourage manufacturers to allocate more resources, labor and capital plant to production, helping to meet additional demand, while falling prices make other uses of resources, labor and capital plant more lucrative and curb supply.

Negative feedback loops of a very similar kind control the production of primary goods by the Earth's natural systems. Every primary good from the water levels in a river and the fertility of a given patch of soil, to more specialized examples such as the pollination services provided by bees to agricultural crops, is regulated by delicately balanced processes of negative feedback working through some subset of the planetary biosphere. The parallel is close enough that ecologists have drawn on metaphors from economics to make sense of their field, and it's quite possible that an ecological economics using natural systems as metaphors for the secondary economy could return the favor and create an economics that makes sense in the real world.

It's when we get to the tertiary economy of financial goods that things change because the feedback loops governing tertiary goods are not negative but positive. Imagine a thermostat designed by a sadist. In the summer, whenever the temperature goes up above a certain level, the sadothermostat makes the heat come on and the house gets even hotter; in the winter, when the temperature goes below another threshold, the temperature shuts off and the house gets so cold the pipes freeze. That's positive feedback, and it's the way the tertiary economy works when it's not constrained by limits imposed by the primary or secondary economies.

The recent housing bubble is a case in point. It's a remarkable case, not least because houses — which are usually part of the secondary economy, being tangible goods created by human labor — were briefly and disastrously converted into tertiary goods, whose value consisted primarily in the implied promise that they could be cashed in for a higher price at some future time. (As a tertiary good, their physical structure had no more to do with their value than does the paper used to print a bond.) When the price of a secondary good goes up, demand decreases, but this is not what happened in the housing bubble; instead, the demand increased, since rising prices made further appreciation appear more likely, and the assorted mis-, mal- and nonfeasance of banks and mortgage companies willing to make seven-figure loans to anyone who could fog a mirror removed all limits from the supply.

The limits, rather, were on the demand side, as they always are in a speculative bubble. Eventually the supply of buyers runs out, because everyone who is willing to plunge into the bubble has already done so. Once this happened to the housing bubble, prices began to sink, and once again positive feedback came into play as the sadothermostat shifted from heating an overheated house to cooling an overchilled one. Since the sole value of these homes to most purchasers consisted of the implied promise that they could be cashed in someday for more than their purchasers paid for them, each decline in price convinced more people that this would not happen, and drove waves of selling that forced the price down further. This process typically bottoms out when prices are as far below the median as they were above it at the peak, and for a similar reason: as a demand-limited process, a speculative bubble peaks when everyone willing to buy has bought, and bottoms when everyone capable of selling has sold.

It's important to note that in this case, as in many others, the positive feedback in the tertiary economy disrupted the workings of the secondary economy. Long before the housing boom came to its messy and inevitable end, there was a massive oversupply of housing in many markets—there are, for example, well over 50,000 empty houses in Phoenix, Arizona, as I write these words, most of them manufactured during the boom for sale to speculators. Absent a speculative bubble, the mismatch between supply and demand would have brought the production of new houses to a gentle halt. Instead, due to the positive feedback of the tertiary economy, supply massively overshot demand, leading to a drastic misallocation of resources in the secondary economy, and thus to an equally massive recession.

It's long been popular to compare the tertiary economy to gambling, but the role of positive feedback in the tertiary economy introduces an instructive difference. When four poker players sit down at a table and the cards come out, their game has negative feedback. The limiting factor is the ability of the players to make good on their bets; the amount of wealth in play at the start of the game is exactly equal to the amount at the end, though it's likely to go through quite a bit of redistribution. For every winner, in other words, there is an equal and opposite loser.

The tertiary economy does not work this way. When a market is going up, everyone invested in it gains; when it goes down, everyone invested in it loses. Paper wealth appears out of thin air on the way up, and vanishes into thin air on the way down. The difference between this and the supply-limited negative feedback cycles of the natural environment could not be more marked. In this sense it's not unreasonable to call the tertiary economy a kind of anti-ecology, a system in which all the laws of ecology are stood on their heads — until, that is, the delusional patterns of behavior generated by the tertiary economy collide with the hard limits of ecological reality.

It's not all that controversial to describe financial bubbles in this way, though you can safely bet that during any given bubble, a bumper crop of economists will spring up to insist that the bubble isn't a bubble and that rising prices for whatever the speculation du jour happens to be are perfectly justified by future prospects. On the other hand, it's very controversial just now to suggest that the entire tertiary economy is driven by positive feedback. Still, I suggest that this is a fair assessment of the financial economy of the industrial world, and the only reason that it's controversial is simply that we, our great-grandparents' great-grandparents, and all the generations in between have lived during the upward arc of the mother of all speculative bubbles.

The vehicle for that bubble has not been stocks, bonds, real estate, derivatives or what have you, but industrialism itself: the entire project of increasing the production of goods and services to historically unprecedented levels by amplifying human labor with energy drawn from the natural world, first from wind and water, and then from fossil fuels in ever-increasing amounts. Like the real estate at the core of the recent boom and bust, this project had its roots in the secondary economy, but quickly got transformed into a vehicle for the tertiary economy: people invested their money in industrial projects because of the promise of more money later on.

Like every other speculative bubble, the megabubble of industrialism paid off spectacularly along its upward arc. It's inaccurate to claim, as some of its cheerleaders have, that everybody benefited from it; one important consequence of the industrial system was a massive distortion of patterns of exchange in favor of the major industrial nations, to the massive detriment of the rest of the planet. (It's rarely understood just how much of today's Third World poverty is a modern phenomenon, the mirror image and necessary product of the soaring prosperity of the industrial nations.)[10] Still, for some three hundred years, standards of living across the industrial world soared so high that people of relatively modest means in America or western Europe had access to goods and services not even emperors could command a few centuries before.

In the absence of ecological limits, it's conceivable that such a process could have continued until demand was exhausted, and then unraveled in the usual way. The joker in the deck, though, was the dependence of the industrial project on the extraction of fossil fuels at an ever-increasing pace. Beneath the giddy surface of industrialism's bubble, in other words, lay the hard reality of the tertiary economy's dependence on the productive capacity of the secondary economy, which is itself dependent on resources from the primary economy. The positive feedback loop driving the industrial bubble can't make resources out of thin air — only money can be invented so casually — but it has proven quite successful at preventing industrial economies from responding to the depletion of their fossil fuel supplies fast enough to stave off what promises to be the great-grandmother of all speculative busts. The results of this failure are beginning to come home to roost in our own time.

The Finance Trap

To understand how this works, it's necessary to take a closer look at the impact of energy depletion on economic life. The arrival of geological limits to increasing fossil fuel production places a burden on the secondary economy, because the cost — measured in energy, labor and materials, rather than money — to extract fossil fuels does not depend on market forces. On average, that cost increases steadily as easily accessible reserves are depleted and have to be replaced by those more difficult and costly to extract. Improved efficiencies and new technologies can counter that to a limited extent, but both these face the familiar problem of diminishing returns as the laws of thermodynamics come into play.

As a society nears the geological limits to production, a steadily growing fraction of its total supplies of energy, resources and labor have to be devoted to the task of bringing in the energy that keeps the entire economy moving, and must therefore be diverted from other economic purposes. This percentage may be small at first, but it functions as a tax in kind on every productive economic activity, and as it grows it makes productive economic activity less profitable. The process by which money produces more money consumes next to no energy, by contrast, and so financial investments don't suffer from rising energy costs to anything like the same extent.

This makes investing in the tertiary economy, on average, relatively more profitable than investing in the secondary economy of nonfinancial goods and services. The higher the burden imposed by energy costs, the more sweeping the disparity becomes. The result, of course, is that individuals trying to maximize their own economic gains move their money out of investments in the productive economy of goods and services and into the paper economy of finance.

Ironically, this is happening just as a perpetually expanding money supply driven by mass borrowing at interest has become an anachronism unsuited to the new economic reality of energy contraction. It also guarantees that any attempt to limit the financial sphere of the economy will face mass opposition, not only from financiers but from millions of ordinary citizens whose dreams of modest wealth and a comfortable retirement depend on the hope that financial investments will outperform the faltering economy of goods and services. Meanwhile, just as the economy most needs massive reinvestment in productive capacity to retool itself for the very different world defined by contracting energy supplies, investment money seeking higher returns flees the secondary economy for the realm of tertiary paper wealth.

Nor will this effect automatically be countered by a flood of investment money going into energy production and bringing the cost of energy back down. Producing energy takes energy, and thus is just as subject to rising energy costs as any other productive activity; as the price of oil goes up, the costs of extracting it or making some substitute for it rise in tandem and make investments in oil production or replacement no more lucrative than any other part of the secondary economy. Oil that has already been extracted from the ground may be a good investment, and financial paper speculating on the future price of oil will likely be an excellent one, but neither of these help increase the supply of oil or any oil substitute.

One intriguing detail of this scenario is that it has already affected the first major oil producer to pass the peak of its oil production, the United States. It's no accident that in the wake of its own 1972 production peak, the American economy has followed exactly this trajectory of massive disinvestment in the productive economy and massive expansion of the paper economy of finance. Plenty of other factors played a role in that process, no doubt, but I suspect that the erratic but inexorable rise in energy costs over the last forty years or so may have had much more to do with the gutting of the American economy than most people suspect.

Now that petroleum production has encountered the same limits globally that put it into a decline here in the United States, the same pattern of disinvestment in the secondary economy coupled with metastatic expansion of the tertiary economy of money is showing up on a much broader scale. There are limits to how far it can go, of course, not least because financiers and retirees alike are fond of consumer goods now and then, but those limits have not been reached yet, not by a long shot. It's all too easy to foresee a future in which industry, agriculture and every other sector of the secondary economy suffer from chronic underinvestment, energy costs continue rising and collapsing infrastructure becomes a dominant factor in daily life, while the Wall Street Journal (perhaps printed in Shanghai by then) announces the emergence of the first half dozen quadrillionaires in the derivatives-of-derivatives-of-derivatives market.

Perhaps the most important limit in the way of such a rush toward economic absurdity is the simple fact that not every economy uses the individual decisions of investors pursuing private gain to allocate investment capital. It is not accidental that quite a few of the world's most successful economies just now — China comes to mind — make their investment decisions at least in part on political, military and strategic grounds, while the nation that preens itself most proudly on leaving the market to distribute investment funds, the United States, lurches from one economic debacle to another.

Those debacles follow a pattern that, at this point, should be utterly familiar to everyone. Beginning in the wake of the harrowing 1987 Wall Street crash, in which the market shed close to a quarter of its total value in a single day, the US government embarked on a policy of dealing with economic crises by dropping interest rates and flooding markets with plenty of cheap debt. The low interest rates and abundant money this strategy produced sparked a series of speculative bubbles; as these caused financial crises of their own, the government repeated the exercise on progressively larger scales.

This habit is not unique to the United States. Japan did exactly the same thing in the wake of its stock market and real estate crash of 1990, Britain has done it during its own recent economic crises and the European Union became the latest member of the club in 2010 when its central bank flooded the Eurozone with cheap debt in an attempt to deal with the near-bankruptcy of Greece. These days, the entire industrial world is drowning in excess debt, and when this causes another bubble and bust — as it inevitably will — the answer of choice will doubtless be the manufacture of even more tertiary wealth by way of another round of borrowing.

It's harsh but not, I think, unfair to characterize this strategy as trying to put out a house fire by throwing buckets of gasoline onto the blaze. Still, a complex history and an even more complex set of misunderstandings feed this particular folly. Nobody in the corridors of power has forgotten what happened the last time a major depression was allowed to run its course unchecked by government manipulation, and every industrial nation has its neofascist fringe parties who are eager to play their assigned roles in a remake of that ghastly drama. That's the subtext behind the international effort to talk tough about austerity while doing as little as possible to make it happen, and the even wider effort to game the global financial system so that Europe and America can

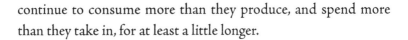

continue to consume more than they produce, and spend more than they take in, for at least a little longer.

Before Money

A primary reason why these sweeping changes have been able to happen unnoticed by economists and policy makers is that today's economics focuses almost entirely on the fit between the secondary and tertiary economies. There's some justification for this focus, for these two can easily get out of step with each other, and the resulting mismatch can cause serious problems. When there's more money in circulation than there are goods and services for the money to buy, you get inflation; when the mismatch goes the other way, you get deflation; when the mechanisms that provide credit to business enterprises gum up, for any number of reasons, you get a credit crunch and recession, and so on. In extreme cases, which used to happen fairly often until the aftermath of the Great Depression pointed out what the cost could be, several of these mismatches could hit at once, leaving both the secondary and tertiary economies crippled for years at a time.

This is the sort of thing that conventional economic policy is meant to confront, by fiddling with the tertiary economy to bring it back into balance with the secondary economy. The industrial world hasn't had a really major depression since the end of the 1930s because the methods cobbled together by governments to fiddle with the tertiary economy work tolerably well. It's become popular in recent years to insist that the unfettered free market is uniquely able to manage economic affairs in the best possible way, but such claims fly in the face of all the evidence of history. The late nineteenth century, for example, when the free market was as unfettered as it's possible for a market to get, saw catastrophic booms and busts sweep through the industrial world with brutal regularity, causing massive disruption to economies around the world.[11] Those who think this is a better state of affairs than the muted ebbs and flows of the second half of the twentieth century are welcome to try living in a Depression-era tarpaper shack on a dollar a day for a week or two.

The problem we face now is that the arrangements evolved over the last century or so only address the relationship between the secondary and tertiary economies. The primary economy of Nature, the base of the entire structure, is ignored by most contemporary economists, and has essentially no place in the economic policy of today's industrial nations. The assumption hardwired into nearly all modern thought is that the economic contributions of the primary economy will always be there so long as the secondary and tertiary economy are working as they should. This may just be the Achilles' heel of the entire structure, because it means that mismatches between the primary economy and the other two economies not only won't be addressed—they won't even be noticed. The resulting blindness to the foundations of our economic life transforms the economic superstitions discussed in Chapter One from a minor nuisance to a potentially fatal liability.

The rising curve of economic volatility over the last decade or so thus marks the point where the primary economy of Nature will no longer support the standard of living most people in the industrial world expect. Our politicians and economists are trying to deal with the resulting crises as though they were purely a product of mismatches between the secondary and tertiary economies. Since such measures don't address the real driving forces behind the crises, they fail, or at best stave off trouble for a short time at the expense of making it worse later on.

The signals warning us that we have overshot the capacity of the primary economy are all around us. The peaking of world conventional oil production in 2005 is only one of these. The widespread dieoff of honeybees is another, on a different scale; whatever its cause, it serves notice that something has gone very wrong with one of the natural systems on which human production of goods and services depends. It's easy to dismiss any of these signals individually as irrelevancies, but every one of them has an economic cost, and every one of them serves notice that the natural systems that make human economic activity possible are cracking under the strain we've placed on them.

That prospect is daunting enough. There's another side to our predicament, though, because the only tools governments know how to use in response to economic trouble are ways of fiddling with the tertiary economy. When those tools don't work — and these days, increasingly, they don't — the only option policy makers seem to be able to think of is to do more of the same, following what's been called the "lottle" principle — "if a little doesn't work, maybe a lot'll do the trick." The insidious result is that the tertiary economy of money is moving ever further out of step with the secondary economy of goods and services, yielding a second helping of economic trouble on top of the one already dished out by the damaged primary economy. Flooding the markets with cheap credit may be a workable strategy when a credit crunch has hamstrung the secondary economy; when what's hitting the secondary economy is the unrecognized costs of ecological overshoot, though, flooding the markets with cheap credit simply accelerates economic imbalances that are already battering economies around the world.

One interesting feature of this sort of two-sided crisis is that it's not a unique experience. Most of the past civilizations that overshot the ecological systems that supported them, and crashed to ruin as a result, backed themselves into a similar corner. I've mentioned already the way that the classic Lowland Mayans tried to respond to the failure of their agricultural system by accelerating the building programs central to their religious and political lives. Their pyramids of stone served the same purpose as our pyramids of debt: they systematized the distribution of labor and material wealth in a way that supported the social structure of the Lowland Mayan city-states and the *ahauob* or "divine kings" who ruled them. Yet building more pyramids was not an effective response to topsoil loss; in fact, it worsened the situation considerably by using up labor that might have gone into alternative means of food production.

An even better example, because a closer parallel to the present instance, is the twilight of the Roman world. Ancient Rome had a sophisticated economic system in which credit and government stimulus programs played an important role. Roman money, though, was based strictly on precious metals, and the economic expansion of the late republic and early empire was made possible only because Roman armies systematically looted the wealth of most of the known world. More fatal still was the shift that replaced a sustainable village agriculture across most of the Roman world with huge slave-worked *latifundiae*, the industrial farms of their day, which were treated as cash cows by absentee owners and, in due time, were milked dry. The primary economy cracked as topsoil loss caused Roman agriculture to fail; attempts by emperors to remedy the situation failed in turn, and the Roman government was reduced to debasing the coinage in an attempt to meet a rising spiral of military costs driven by civil wars and barbarian invasions. This made a bad situation worse, gutting the Roman economy and making the collapse of the empire that much more inevitable.

It's interesting to note the aftermath. In the wake of Rome's fall, lending money at interest — a normal business practice throughout the Roman world — came to a dead stop for centuries. Christianity and Islam, which became the majority religions across what had been the empire's territory, defined it as a deadly sin. More, money itself came to play an extremely limited role in large parts of the former Empire. Across Europe in the early Middle Ages, as already mentioned, it was common for people to go from one year to the next without so much as handling a coin. What replaced money was the use of labor as the basic medium of exchange. That was the foundation of the feudal system, from top to bottom: from the peasant who held his small plot of farmland by providing a fixed number of days of labor each year in the local baron's fields, to the baron who held his fief by providing his overlord with military service, the entire system was a network of personal relationships backed by exchanges of labor for land.

It's common in contemporary economic history to see this as a giant step backward, but there's good reason to think it was nothing of the kind. The tertiary economy of the late Roman world had become a corrupt, metastatic mess; the new economy of feudal Europe responded to this by erasing the tertiary economy as completely as possible, banishing economic abstractions and producing a stable and resilient system that was very hard to game — deliberately failing to meet one's feudal obligations was the one unforgivable crime in medieval society, and the penalty usually involved the prompt and heavily armed arrival of one's liege lord at the head of all his other vassals.

What makes this even more worth noting is that very similar systems emerged in the wake of other collapses of civilizations. The implosion of Heian Japan in the tenth century, to name only one example, gave rise to a feudal system so closely parallel to the European model that it's possible to translate much of the technical language of Japanese bushido precisely into the equivalent jargon of European chivalry and vice versa. More broadly, when complex civilizations fall apart, one of the standard results is the replacement of complex tertiary economies with radically simplified systems that do away with tertiary abstractions such as money, and replace them with the concrete secondary economics of land and labor.

In the last chapter of this book, we'll discuss ways that this might be set in motion that don't necessarily involve the disintegration of society in the kind of maelstrom of violence and fragmentation that set the feudal systems of Europe and Japan in motion. Before we get there, however, it's necessary to take a closer look at money — the central element of our civilization's tertiary economy — and recognize the complex and ultimately metaphysical assumptions that are quietly transforming money from a system for distributing goods and services into a weapon of mass destruction that may well bring our civilization to its knees.

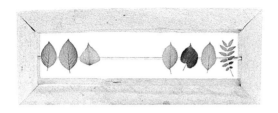

THE METAPHYSICS *of* MONEY

T O MENTION MONEY and metaphysics in the same sentence is to invite any number of misunderstandings. The hoary habit of thinking that walls off philosophical questions in a ghetto of abstractions apart from the world of ordinary life gets in the way of clarity here, as so often, but there's an even more basic problem: most people these days have no clear notion of what the word "metaphysics" means in the first place.

The tangled history of the word probably makes that inevitable. A nameless librarian in ancient Alexandria first coined it out of sheer desperation while cataloging the works of Aristotle; most of the treatises got names based on their subject matter — *Physics, Meteorology, Poetics*, and so on — but one difficult treatise was labeled simply *meta phusikoi*, "the stuff that comes after the *Physics*." Then, as the famous fourth-grade history paper put it, some other stuff happened — the library of Alexandria burned, Rome fell, what was left of the classical world got tipped into history's dumpster by a band of helpful Visigoths and so on. When the dust finally cleared, Aristotle was very nearly the only systematic ancient thinker whose works were still available in the Western world, and so he became, in Dante's words, "the master of those who know."

That meant, among other things, that the labels assigned to his treatises by that anonymous Alexandrian savant became the basic categories of scholarship in the Middle Ages. (Most of them remain basic categories today, which is why your local university has departments of physics, meteorology and so on.) Metaphysics was no exception, and the philosophical issues Aristotle tackled in that treatise have carried that label ever since.

Those issues are what Aristotle himself called "first philosophy": an analysis of the basic terms that have to be sorted out before any kind of philosophy can be sure of its foundations. The medieval scholars who blew the dust off Aristotle's treatise, however, interpreted his work in their own way, which meant that the basic issues of philosophy were redefined in terms of Christian, Muslim or Jewish theology. By the time the eighteenth century rolled around, metaphysics as a discipline was almost entirely identified with the theological basis given it by the scholars of the Middle Ages, and so it got dropped like a hot potato as secularism swept the academic world.

By the end of the nineteenth century even theologians had stopped doing metaphysics in the old style, and most of the people practicing what used to be called metaphysics weren't using the word. At that point, in a fine display of history's twisted sense of humor, the word got picked up by the American folk religious movement ancestral to today's New Age scene, and turned into a label for their own beliefs. The town in southern Oregon where I used to live has a Metaphysical Library, which even had a few books on metaphysics in the philosophical sense of the world, though how they got there I have no idea. The vast majority of the books were on past lives, channeled entities, flying saucers, evil conspiracies and the rest of the mental furniture of contemporary alternative culture.

Thus it's probably necessary to point out that when I mention the metaphysics of money, I'm not referring to claims that money was invented by a conspiracy of evil space lizards, or that you can get as much money as you want by convincing yourself that money really, really loves you and wants to bed down in your wallet. You can find books making both these claims at the library just mentioned, as it happens, but both beliefs — and a good many statements less obviously absurd — are in large part produced by a failure to engage in the other kind of metaphysics, the thoughtful consideration of the basic categories of thought itself.

That sort of analysis seems very abstruse and impractical, until you notice the consequences of ignoring it. Sweeping claims are being made these days about whether certain things exist or do not exist, for example, by people who never seem to have examined their own presuppositions about what it means to exist and how a thing can be known to exist. The ongoing squabble between atheists of the Richard Dawkins stripe and their religious critics is perhaps the best known example of this; both sides are talking past each other because neither has grappled with the implications of its own metaphysical claims. That's the problem with the ghettoizing of philosophy mentioned earlier: the philosophical issues you ignore can still sneak up on you while you're not looking, and turn your best attempts at thinking into gibberish.

The point of contact between metaphysics and money is precisely the need to be sure of the meaning of basic concepts too often taken for granted. There are several reasons why you can reliably get better economic advice from a randomly chosen fortune cookie at your local Asian buffet than from the most prestigious contemporary economists. As Chapter One pointed out, the fact that offering bad advice very often becomes a ticket to career success in economics is certainly a factor, and so are the influences discussed a little later in the same chapter — the assumption that economics can be understood apart from the non-economic factors that shape them, the faith in the existence and omnipotence of free markets, the insidious distortions caused by faked statistics and the impact of paradigms and superstitions.

Chapter One also mentioned the impact of premature mathematic formulations on economics, and that habit requires a second look at this point. As a barrier to clear thought, it's far from unique to economics — in one way or another, it underlies a great many of the mistakes that are tipping our own civilization toward the same dumpster that received the ruins of Rome — but it stands out in the field of economics with particular clarity. Its roots are in a metaphysical error which might as well be called, after one of its most influential practitioners, Descartes' fallacy.

René Descartes is famous nowadays for saying "I think, therefore I am." Few people these days take the time to find out what he meant by that statement, and fewer still catch onto the radical project that underlay it. Without too much inaccuracy, Descartes can be called the first modern thinker. Certainly he was the first to embrace what has become an automatic presupposition of modern thought, the notion of the individual self as an isolated, independent witness whose thoughts and experiences are entirely its own. What existed, to Descartes, was limited to what he could know, and know precisely, with the same exactness as a geometrical proof.

Descartes was arguing, in effect, that "to be" means the same thing as "to be known," and "to be known" in turn equals "to be precisely defined." It's clear that he recognized, and intended, the sweeping implications of this metaphysical stance. It's equally clear that a great many of the people who unknowingly follow his lead nowadays either accept those implications uncritically or have never noticed their existence. In the hands of much of modern science, in particular, Descartes' equation has been blended with a passion for quantitative measurement to produce an even more extreme form of the same logic. To a great many scientists today, what exists is limited to what can be known; what can be known is limited to what can be measured; and what can be measured is treated as though it was identical to its measurements.

You can get away with this in physics and still do excellent science. The objects studied by physics follow patterns that can be modeled effectively by mathematics, and most of them are so remote from ordinary human experience that anything about them that can't be easily measured can be ignored without too much trouble. Try doing this in sciences closer to the realm of everyday human life, on the other hand, and you can count on running into trouble, because in that realm Descartes' approach is usually a bad idea, and the modern scientific expansion of it an even worse one. What can be measured is only a subset of what can be known, and what can be known, at least in any given situation, is only a subset of what exists; nor does the fact that some properties of a thing can be measured according to some numerical scale prevent it from having other properties at least as important that are not subject to that kind of measurement.

The sort of bad logic that treats quantitative measurements as the only things that really exist is pervasive in the sciences, but its grip is even tighter on those fields of study that want to claim the prestige of science but can't quite pass muster. Economics is the poster child for this noxious effect. Down through the generations, against the sound advice of its best practitioners, economists have consistently treated the one thing in their field that can easily and consistently be measured with numbers — money — as though it was the one thing that matters. It's easy to see how seductive this habit can be, since it seems to allow everything to be measured on a common scale; the problem, of course, is that everything that can't be flattened out into that common scale gets mislaid, and as often as not these mislaid factors prove to be decisive.

In *The Wealth of Nations*, Adam Smith criticizes the notion — as common in his time as in ours — that money is the same thing as wealth.[1] The wealth of a country, he points out, consists of the product of its natural resources and collective labor: in the terms I've sketched out in the previous chapter, it's the sum total of the goods and services produced by a nation's primary and secondary economies. In another place, though, he defines wealth as anything that can be valued in money. These definitions do not conflict with one another; rather, they make the crucial point that money is not wealth but *the yardstick by which modern cultures measure wealth*. This ought to be the first thing we teach children about money, though, of course, it isn't.

It probably ought to be the first thing we teach economists about money, too, but the power of Descartes' fallacy stands in the way. Money is a unit of measurement, so it's inherently easy to define, understand and quantify. Real wealth is much less easy to force into the Procrustean bed of numbers; that's why we use money as a rough and ready way of sorting out the relative value of different kinds of wealth so they can be exchanged without too much trouble. Money is so convenient as a way of measuring wealth that very often it ends up eclipsing wealth, and this is why most economists nowadays, even when they think they're talking about wealth, are actually talking about money. This becomes especially problematic when, as so often happens, they start attributing to wealth characteristics that are only true of money.

This habit of thought pervades contemporary economics. For a crucial example, already mentioned, watch the way most economists these days brush aside the immense challenges of peak oil with the assurance that if oil ever does get scarce, the market will come up with alternatives. Implicit in this claim is the assumption that any energy source is as good as any other, and that the total amount in the system is effectively unlimited. This is true of money — at least in theory, one dollar bill is worth exactly the same amount as any other, and the total number of dollars in circulation is as close to limitless, these days, as the printing presses of the US Treasury can make it — but it is emphatically not true of energy resources, or of any other form of wealth.

Compare any two energy resources in practical terms and it's clear that in most cases they're not even apples and oranges — they're apples and orangutans. Take petroleum and solar energy as good examples. A highly concentrated form of chemical energy and a rather diffuse form of electromagnetic energy have very little in common, and even when they can do the same things — you can heat a house with passive solar design, for example, or you can heat it with an oil-fired burner — the technologies are totally different. Easy talk about swapping one for the other thus evades the immense challenge and nearly unimaginable cost of scrapping multiple continent-wide infrastructures geared to oil and building new ones suited to solar energy. (There are plenty of other questions ducked by talk of that kind, but this one will do for now.)

Presumably an economist would notice something odd if he sat down at a lunch counter, ordered the daily special, and was handed instead a box of socket wrenches, even if the price of the wrenches was exactly the same as the daily special. If the economist was starving on a desert island and a crate that washed ashore proved to contain socket wrenches rather than food, the difference would be a matter of life or death. This latter is uncomfortably close to our position just now, as the world's energy companies race each other and the clock to extract fossil fuels in nearly unimaginable volumes from the Earth's dwindling supplies. If we allow ourselves to wait until actual energy shortages begin to cripple our capacity to produce goods and services, it will be much too late to start retooling our civilization for some other energy resource, even if one happens to turn up.

Because a subculture of erudite scholars in the economics departments of universities have made a metaphysical error — treating quantitative measurements as the only things that really exist — our civilization may have missed its chance to dodge disaster. It's hard to think of a better argument for the importance of metaphysics than that. Still, the problem sketched here extends much further, and the way in which money has metastasized in our society to become the measure of all things has become a massive though unrecognized barrier in the way of any attempt to improve a rapidly worsening situation.

The Metastasis of Money

Perhaps the most important force behind the confusion of money and wealth is the way, discussed in Chapter Two, that money has metastasized so deeply into our economic life that it's nearly impossible to do much of anything without it. The economic textbooks you did your best not to read in school justify that ubiquity by a neat rhetorical trick. If you remember anything at all about that, it's probably that bit of rhetoric; it can be found in the canned explanation for why we use money, somewhere around page six. It runs something like this: there's a plumber and a pig farmer who want to do business with one another, see, but the plumber's Jewish and the pig farmer has nothing to trade but pork. Add money, and voila! The farmer sells his pork to other people and uses the proceeds to pay the plumber, who uses it to buy gefilte fish and matzoh meal. Everyone's happy except, presumably, the pigs.

It all seems very logical until you think about it for ten seconds. Notice, to start with, how the explanation assumes that the plumber, the pig farmer, the purchasers of pork, the kosher deli and everyone else are restricted to the specific kind of economic relationships that exist in, and only in, a money economy. None of them have access to any other means of exchange except barter; household economies, gift economies and customary exchange economies—which between them accounted for most economic activity during most of human history—are excluded from the story.

Thus the plumber doesn't, as most people did a century ago, benefit from a household economy that provides a great deal of his food, including small livestock in the back garden. The pig farmer doesn't, as most people did until as little as fifty years ago, do es-

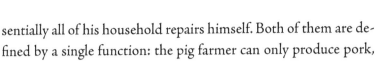

sentially all of his household repairs himself. Both of them are defined by a single function: the pig farmer can only produce pork, the plumber only plumbing. Nor do the farmer, the plumber or anyone else have access to any of the immense variety of nonmonetary systems of exchange human beings have used throughout history.

!Kung hunter-gatherers sharing out a wildebeest among band members according to traditional rules, Haida chiefs distributing blankets and salmon to all comers at a potlatch and medieval peasants working a baron's demesne lands for a set number of days each year to maintain their ownership of their own cottages and fields, all participated in flexible and effective systems of exchange that had nothing to do with money. Urban societies as complex as ancient Egypt got by entirely without money and still managed to keep plumbers, pig farmers and a great many other occupational specialties gainfully employed for millennia.

All that the textbook explanation proves, in other words, is that if you have a money economy, it probably does need some kind of money to make it work. This is not the conclusion the textbooks draw from the plumber and the pig farmer, of course; with very few exceptions, they leap from their canned example to the claim that money must be essential to any economy worth the name, and the rest of the textbook proceeds to focus on theories about the behavior of money under the false impression that those theories deal with the behavior of wealth.

The mistaken metaphysics discussed above plays a large role in fostering this misunderstanding, but the sheer pervasiveness of money in today's industrial economy also has a very significant role. For most people in the modern industrial world, the only way to get access to any kind of wealth — that is, any good or service — is to get access to money first, and exchange the money for the wealth. This makes it all too easy to confuse money with wealth, and it also fosters the habit of thought that treats money as the driving force in economic life, and thinks of wealth as a product of money, rather than seeing money as an arbitrary measure of wealth.

The thought experiment already mentioned of placing a plane-load of economists on a desert island with one million dollars each but no food or water is a good corrective to this delusion. Unfortunately, this same experiment is being tried on a much vaster scale by the world's industrial economies right now. We have seven billion people on a planet with a finite and dwindling supply of the concentrated energy resources that are keeping most of them alive, and governments and businesses alike are acting as though the only possible difficulty in this situation is coming up with enough money to pay for investments in the energy industry.

It should be obvious that no amount of money can overcome the natural laws that have placed hard limits on the amount of highly concentrated energy resources that happen to exist on our planet. This is not obvious to most people nowadays, however, because the metastasis of money throughout the economy has trained nearly all of us to think that if you have enough money you can get whatever you want. The fact that the richest people in the world can put their entire fortunes into health care and still get old and die is one of the few persistent reminders that money cannot overcome the laws of Nature or provide access to goods and services that don't exist.

So how did money get transformed from a convenient yardstick for real wealth to the be-all and end-all of contemporary economic life? At least three factors were involved, two of them common to complex urban societies throughout history, one unique to ours.

First, despite the drastic oversimplifications of the textbook example cited earlier, it reflects a reality: a complex society can gain significant advantages from a medium of exchange that can be traded for any form of wealth. Even in societies where most goods and services are distributed by way of social networks, social consensus tends to establish certain trade goods — wampum shell strings among the First Nations of eastern North America, for instance — as a common measure for those goods and services that are exchanged in other ways. As a society becomes more complex and the division of labor among different crafts expands, some standard measure of wealth becomes more useful. While money itself was invented around 700 BCE by the ancient Greeks, other ways of measuring wealth for the sake of easy exchange had been in use in Old World urban societies for millennia before then.

Second, whenever common measures of wealth are controlled by institutions, those who manage those institutions become powerful and can be counted on to maintain and expand their power whenever possible. In ancient Egypt, for example, grain in temple warehouses provided the basic measure of wealth; as a result, the priests who controlled the stockpiled grain became a potent political force. In medieval Europe, when land was the basic measure of wealth — there's a reason we still call land "real estate," as though all other wealth is unreal — the power of the feudal nobility derived directly from their control of land. Today the governments that claim exclusive power to print and regulate money, and the banks and financial corporations that manage most of society's money, derive much of their effective power from their control over the medium of economic exchange, and can be counted on to encourage the rest of society to rely ever more completely on the thing that gives them power.

These two factors can be traced in the history of most of the complex urban societies of the past. What makes our civilization something of an extreme case is a third factor — the extreme complexity of an economic system that has temporarily replaced the limited energy resources of other human societies with a torrent of cheap and abundant energy from fossil fuels.

Ilya Prigogine, one of the most innovative physicists of recent years, showed via a series of dizzyingly complex equations that the flow of energy through a system increases the complexity of the system.[2] Any doubt of the accuracy of his claim can be settled by regarding the economic history of the Western world from 1700 to the present. The societies over which the tsunami of the Industrial Revolution broke in the early eighteenth century were not unusually complex by the standards of past civilizations; their own contemporaries in the Chinese and Ottoman Empires considered western Europeans, not without reason, to be grunting, smelly barbarians with few of the arts and graces of civilization.

Fossil fuels may not have done anything about the gracelessness and the smell, but it certainly made up for any shortage in complexity. Until the dawn of the industrial age, as a general rule of thumb, some 90 percent of the inhabitants of any complex society worked in agriculture, providing the food and raw materials that supported themselves as well as the 10 percent who could be spared for all other economic roles. By 1900, at the zenith of the age of coal, many nations in the industrial world had dropped the percentage of their work force in agriculture below 50 percent, and shifted the workers thus freed up into a broad assortment of new economic roles. By 2000, buoyed by the much higher concentration and efficiency of petroleum, many industrial nations had dropped the percentage of their work force in agriculture below 5 percent, with the other 95 percent filling newly invented roles in the most complex economies in the history of the planet.

One consequence of this swift and unprecedented surge in complexity was the triumph of money over all other systems of exchange. When the vast majority of workers at every income level labored at tasks so specialized that their efforts only produced value when combined with those of hundreds or thousands of other workers, money provided the only way they could receive a return on their labor. When most of the customers for any given product had money and nothing else to exchange for it, buying products for money became standard. Social networks of exchange—household economies, customary local exchanges, church and fraternal networks—shattered under the strain, and were replaced by purely economic relationships—wage labor, shopping, public assistance—that could be denominated entirely in cash. The last three centuries of social and economic history are largely a chronicle of the results.

If economists took a wider view of the history of their discipline than they generally do, they might have noticed that what most of them consider a fundamental feature of all economies worth studying—the centrality of money—is actually a unique feature of an economic era defined by unprecedented amounts of cheap energy. Since the fossil fuels that made that era possible are being extracted at a pace many times the rate at which new supplies are being discovered, current assumptions about the role of money in society may be in for a series of unexpected revisions.

In an ironic way, this process of revision may be fostered by the antics of the world's industrial nations as they try to forestall the Great Recession by spending money they don't have. The economic crisis that gripped the world in 2008 was primarily driven by a drastic mismatch between money and wealth. When the price of a rundown suburban house zoomed from $75,000 to $575,000, for example, the change marked a distortion in the yardstick rather than any actual increase in the wealth being measured. That distortion caused every economic decision based on it—for example, a buyer's willingness to go over his head into debt to buy the house, or a bank's willingness to lend money on the basis of imaginary equity—to suffer similar distortions. Now that the yardsticks have snapped back to something like their proper length, the results of the distortion have to be cleared out of the economy if the amount of money in the system is once again to reflect the actual amount of wealth.

Yet this is exactly what governments and businesses across the industrial world are doing their best to forestall. Governments are scrambling to drive economic activity at a pace the real wealth of their societies can no longer support. Banks and businesses are doing everything in their power to divert attention from the fact that a great many of the financial assets propping up their balance sheets were never worth anything in the first place and now, if possible, are worth even less. Both are doing so by the simple expedient of spending money they don't have. As government deficits worldwide spin out of control and the total notional value of the world's financial paper climbs steadily into the blue sky on the far side of the one quadrillion dollar mark, the decoupling of money from wealth is even more extreme than it was at the height of the real estate bubble. This is another context in which a wider view of history than economists usually allow themselves to take could offer a useful warning.

The dominance of money in complex societies has a distinctive trajectory over time, and that trajectory promises to unfold in ways that will overturn many of the assumptions of conventional economics in the decades ahead. It's ironic to note that this trajectory was first and, in many ways, most usefully traced by a nearly forgotten intellectual who had essentially no interest in the field of economics at all.

The Flight Into Abstraction

Giambattista Vico, who lived from 1668 to 1744, spent his career as a teacher of rhetoric at the University of Naples, and devoted his off hours to one of the great intellectual projects of his time. His masterpiece, *Principles of a New Science Concerning the Common Nature of Nations*, appeared in three editions of increasing complexity, the last one after his death, and was almost completely ignored for most of a century thereafter. When it finally found its audience in the mid-nineteenth century, its influence was profound, and continues to this day.

The *New Science*, as the work is generally known, was nothing less than the first modern attempt to make sense of the laws governing history. Vico was perhaps the first modern Western thinker to recognize the parallel historical trajectories of his own society and that of classical Greece and Rome. Using those as his two test cases, he attempted to sketch out "the course the nations run," the process by which a society rises from barbarism to civilization and falls back to barbarism again. With only two examples to work from, Vico inevitably jumped to many conclusions that don't hold up well in the light of a broader view of world history, but some of his ideas are astonishingly prescient, and his basic intuition — that societies go through broadly similar stages on the way from their initial rise to their final collapse — remains central to any attempt to make sense of history on the grand scale.

Vico's argument is complex and difficult to summarize, but one of its core themes is the role of abstraction. In the early days of a culture, Vico pointed out, a wide range of social phenomena focus entirely on specific concrete realities, but these phenomena evolve toward abstraction over the lifespan of the culture. Law codes start out as lists of rules for specific cases and broaden into statements of principles covering infinite variation in practice; words leave behind concrete meanings — how many people nowadays recall that the verb "understand" once meant literally "to stand under," in the sense of upholding or supporting something? — and take on ever more nuanced meanings; religion begins in the shattering impact of the numinous on individual lives and diffuses into elegant theological notions disconnected from the realities of human experience.

So, too, economics. Vico only gives the economic sphere the briefest of mentions in his work — as a scholar of rhetoric and law, his interests lay elsewhere — but the economic history of the Western world fits his scheme precisely. The cultures that clawed their way back up from the chaos that followed the fall of Rome knew only one form of real wealth, and that was agricultural land. The warrior aristocracies that threw back the last barbarian invasions from Europe and imposed a tenuous peace on their battered societies defined themselves by their landholdings; possession of a "knight's fee" — enough land to support a single armored horseman — was the one requirement of noble status in those days. Money existed in the form of coinage, but it had a tiny economic role, and nearly all goods and services moved through customary patterns of exchange in which market forces had no place.

The waning of the Middle Ages saw the gradual replacement of these customary economies with a new economics of precious metal currency. Feudal tenure, by which farmers held the right to their land in exchange for specific duties defined by tradition, gave way to cash rents, and a significant part of the population moved away from the land to proto-industrial wage labor in the newly expanding cities. This was a step toward abstraction: gold and silver coins replaced fields of grain as the basic definition of wealth and made way for concentrations of economic power far more extreme than anything the Middle Ages had seen.

Further abstractions followed. By the seventeenth century, banks began to issue paper receipts for the gold and silver in their vaults, and these receipts could be exchanged like the coinage that backed them. The invention of the banknote was followed promptly by the practice of printing more banknotes than a bank's gold and silver reserves would cover, on the assumption that most of the notes would never be cashed in for metal. When word of this practice spread, the first bank runs followed. In the same way, companies found they could bring in capital by selling shares of their future earnings; the purchasers of these shares then found that their prices could be bid up or down, and stock speculation was born.

Fast forward a few more centuries, and we arrive at today's global economy, which consists primarily of the buying and selling of abstractions. The concept of wealth, which was once limited to the immediate means of production and then shifted to mean the precious metal markers used to denominate the value of production, has now mutated into arbitrary numbers that can be wished into existence by a few keystrokes. When the US government responded to the implosion of financial firms in 2009 by investing $250 billion in the nation's banks, for example, that money did not have to be pulled out of some bank account in the nation's name, much less extracted from the dwindling productive capacities of America's remaining factories and farms; it was conjured into being by government fiat, in order to replace some even vaster sum of abstract wealth that more or less dissolved into twinkle dust over the preceding weeks.

What makes this pursuit of the abstract so dangerous, of course, is that the abstract value of the tertiary economy is not the same thing as the concrete realities of the primary and secondary economies it once represented: green fields and grain in storehouses; strong muscles and the work they accomplish; or, for that matter, factories, the resources that keep them running and the products that come from them. These are real wealth. The layers of economic abstraction piled atop them are simply complex social games that determine who gets access to how much of this real wealth — and those games can become so complex, and so dysfunctional, that they get in the way of the production of real wealth. The flight into abstraction can proceed so far, in other words, that the abstractions interfere with the concrete realities underlying them.

This possibility became appallingly real during a few weeks in 2009. The overnight interbank loan market — an economic abstraction so arcane that not even economists can explain its function in ordinary English — froze up. Desperation moves by the world's central banks managed to prevent a full-scale collapse, but stock markets worldwide panicked and crashed, erasing trillions of dollars in tertiary wealth in a single week.

A stock market crash, it bears remembering, does not cause crop failures, labor shortages or the destruction of industrial machinery. Its impact is purely on the tertiary economy of abstractions built atop the real primary and secondary wealth of land, labor and industrial plant. Yet that impact can be devastating. In the depths of the last Great Depression, for example, the production of goods shrank to a small fraction of what it had been before the 1929 crash; US production of steel, for example, bottomed out in the summer of 1932 at 12 percent of capacity.[3] There was still plenty of land, plenty of laborers and plenty of machines, not to mention millions of families whose breadwinners would have liked nothing so much as a chance to earn money and buy products; the only thing that could not be made to work was the market where abstractions were bought and sold, and without its help, the real economy ground to a halt.

We are facing the same situation now, and official attempts to stabilize the economy are failing because they focus on the abstractions rather than the realities underlying them. The $250 billion just mentioned, for example, need not have been poured down a Wall Street rat hole; it could, for example, have been used to pay for the rebuilding of America's rail network, with dramatic positive effects that would have resonated throughout the economy. That project would have hired hundreds of thousands of workers across the spectrum of skilled and unskilled trades; locomotives and rolling stock would have had to be built, countless miles of track laid and upgraded, stations repaired or built from scratch, and every dollar spent on all these things would ripple outward through the economy, supporting businesses of every kind and refinancing local banks with deposits rather than loans. Projects of the same kind played a large role in helping many countries in the 1930s begin to pull themselves out of the morass of the last Great Depression.

Instead, the $250 billion was assigned the task of making up for a portion of the imaginary wealth that had already evaporated from the balance sheets of banks. Abstraction has triumphed over economic realities, and the multiple impacts of that failure of imagination will be with us for a long time to come. Yet the richest irony of that process is that, as Vico himself would have been the first to point out, this unparalleled expenditure of abstract wealth has helped drive changes with many parallels in the declining civilizations of the past.

The Twilight of Money

One of the least constructive habits of contemporary thought is its unwillingness to grapple with this last point, and that habit has its roots in the common contemporary insistence on the uniqueness of the modern experience. It's true, of course, that fossil fuels have allowed the world's industrial societies to pursue their follies on a more grandiose scale than any past empire has managed, but the follies themselves closely parallel those of previous societies. A difference of scale does not necessarily equal a difference in kind. Tracking the trajectories of past examples of economic folly is one of the few available sources of guidance if we want to know where the current versions of these phenomena are headed.

The metastasis of money through every aspect of life in the modern industrial world is a good example. While no past society, as far as we know, took this process as far as we have, the replacement of wealth with its own abstract representations is no new thing. As we've seen, Vico pointed out back in the eighteenth century that complex societies move from the concrete to the abstract over their life cycles. This movement toward abstraction has important advantages for complex societies, because abstractions can be deployed with a much smaller investment of resources than it takes to mobilize the concrete realities that back them up. We could have resolved 2008's debate about who should rule the United States the old-fashioned way, for example, by having McCain and Obama call their supporters to arms, march to war and settle the matter in a pitched battle amid a hail of bullets and cannon shot on a fine September day on some Iowa prairie. The cost in lives, money and collateral damage, however, would have been far in excess of those involved in an election. In much the same way, the complexities involved in paying office workers in kind, or even in cash, make an economy of abstractions much less cumbersome for all concerned.

At the same time, there's a trap hidden in the convenience of abstractions: the further you get from the concrete realities, the larger the chance becomes that those concrete realities may not actually be there when needed. History is littered with the corpses of regimes that let their power become so abstract that they could no longer counter a challenge on the fundamental level of raw violence. It's been said of Chinese history, and could be said of that of any other civilization, that its basic rhythm is the tramp of hobnailed boots going up stairs, followed by the whisper of silk slippers going back down. In the same way, economic abstractions keep functioning only so long as actual goods and services exist to be bought and sold, and it's only in the pipe dreams of economists that the abstractions guarantee the presence of the goods and services. Vico argued that this trap is a central driving force behind the decline and fall of civilizations: the movement toward abstraction goes so far that the concrete realities are neglected. In the end the realities trickle away unnoticed, until a shock of some kind strikes the tower of abstractions built atop the void the realities once filled, and the whole structure tumbles to the ground.

We are uncomfortably close to such a possibility just now, especially in our economic affairs. Over the last century, with the assistance of the economic hypercomplexity made possible by fossil fuels, the world's industrial nations have taken the process of economic abstraction further than any previous civilization. On top of the usual levels of abstraction — a commodity used to measure value (gold), receipts that could be exchanged for that commodity (paper money), and promises to pay the receipts (checks and other financial paper) — contemporary societies have built an extraordinary pyramid of additional abstractions. Unlike the pyramids of Egypt, furthermore, this one has its narrow end on the ground, in the realm of actual goods and services, and widens as it goes up.

The consequence of all this pyramid building is that there are not enough goods and services on Earth to equal, at current prices, more than a small percentage of the face value of stocks, bonds, derivatives and other fiscal exotica now in circulation. The vast majority of economic activity in today's world consists purely of exchanges among these representations of representations of representations of wealth. This is why the real economy of goods and services can go into a freefall like the one now under way without having more than a modest impact, so far at least, on an increasingly hallucinatory tertiary economy of fiscal abstractions.

Yet an impact it will have, if the freefall proceeds far enough. This was Vico's point, and it's a possibility that has been taken far too lightly both by the political classes of today's industrial societies and by their critics on either end of the political spectrum. An economy of hallucinated wealth depends utterly on the willingness of all participants to pretend that the hallucinations have real value. When that willingness slackens, the pretense can evaporate in record time. This is how financial bubbles turn into financial panics: the collective fantasy of value that surrounds tulip bulbs, or stocks, or suburban tract housing or any other speculative vehicle, dissolves into a mad rush for the exits. That rush has been peaceful to date; but it need not always be.

The Money Bubble

I've argued that the industrial age is in some sense the ultimate speculative bubble, a three-century-long binge driven by the fantasy of infinite economic growth on a finite planet with even more finite supplies of cheap abundant energy. Still, it's important to realize that this megabubble has spawned a second bubble on an even larger scale. The vehicle for this secondary megabubble is money: more precisely, the entire profusion of abstract representations of wealth that dominate our economic life and have all but smothered the real economy of goods and services, to say nothing of the primary economy of natural systems that keeps all of us alive.

Speculative bubbles are defined in various ways, but classic examples — the 1929 stock binge, say, or the late housing bubble — have certain standard features in common. First, the value of whatever item is at the center of the bubble shows a sustained rise in price not justified by changes in the wider economy or in any concrete value the item might have. A speculative bubble in money functions a bit differently than other bubbles, because the speculative vehicle is also the measure of value; instead of one dollar increasing in value until it's worth two, one dollar *becomes* two. Where stocks or tract houses go zooming up in price when a bubble focuses on them, what climbs in a money bubble is the total amount of paper wealth in circulation. That's certainly happened in recent decades.

A second standard feature of speculative bubbles is that they absorb most of the fictive value they create, rather than spilling it back into the rest of the economy. In a stock bubble, for example, a majority of the money that comes from stock sales goes right back into the market; without this feedback loop, a bubble can't sustain itself for long. In a money bubble, this same rule holds good: most of the paper earnings generated by the bubble end up being reinvested in some other form of paper wealth. Here again, this has certainly happened; the only reason we haven't seen thousand-percent inflation as a result of the vast manufacture of paper wealth in recent decades is that most of it has been used solely to buy even more newly manufactured paper wealth.

A third standard feature of speculative bubbles is that the number of people involved in them climbs steadily as the bubble proceeds. In 1929, the stock market was deluged by amateur investors who had never before bought a share of anything; in 2006, hundreds of thousands, perhaps millions, of people who previously thought of houses only as something to live in came to think of them as a ticket to overnight wealth, and sank their net worth in real estate as a result. The metastasis of the money economy discussed earlier is another example of the same process at work.

Finally, of course, bubbles always pop. When that happens, the speculative vehicle du jour comes crashing back to earth, losing the great majority of its assumed value, and the mass of amateur investors, having lost anything they made and usually a great deal more, trickle away from the market. This has not yet happened to the current money bubble. It might be a good idea to start thinking about what might happen if it does.

The effects of a money panic would be focused uncomfortably close to home, I suspect, because the bulk of the hyperexpansion of money in recent decades has focused on a single currency, the US dollar. That bomb might have been defused if the collapse of the housing bubble in 2008 had been allowed to run its course, because this would have eliminated no small amount of the dollar-denominated abstractions generated by the excesses of recent years. Unfortunately, the US government chose instead to try to reinflate the bubble economy by spending money it didn't have, through an orgy of borrowing and some very dubious gimmickry. A great many foreign governments are accordingly becoming reluctant to lend the US more money, and at least one rising power — China — has been quietly cashing in its dollar reserves for commodities and other forms of far less abstract wealth.

Up until now, it has been in the best interests of other industrial nations to prop up the United States with a steady stream of credit, so that it can bankrupt itself by fulfilling its self-imposed role as global policeman. It's been a very comfortable arrangement, since other nations haven't had to shoulder the costs of dealing with rogue states, keeping the Middle East divided against itself or maintaining economic hegemony over an increasingly restive Third World, while they still received the benefits of all these policies. The end of the age of cheap fossil fuel, however, has thrown a wild card into the game. As world petroleum production falters, it must have occurred to the leaders of other nations that if the United States no longer consumed roughly a quarter of the world's fossil fuel supply, there would be a great deal more for everyone else to share out. The possibility that other nations might decide that this potential gain outweighs the advantages of keeping the United States solvent may make the next decade or so interesting, in the sense of the famous Chinese curse.

Over the longer term, it's safe to assume that the vast majority of paper assets now in circulation, whatever the currency in which they're denominated, will lose all their value. This might happen quickly, or it might unfold over decades, but the world's supply of abstract representations of wealth is so much vaster than its supply of concrete wealth that something has to give sooner or later. Future economic growth won't make up the difference: the end of the age of cheap fossil fuel makes growth in the real economy of goods and services a thing of the past, outside of rare and self-limiting situations. As the limits to growth tighten and become first barriers to growth and then drivers of contraction, shrinkage in the real economy will become the rule, heightening the mismatch between money and wealth and increasing the pressure toward depreciation of the real value of paper assets.

Once again, though, all this has happened before. Just as increasing economic abstraction is a common feature of the growth of complex societies, the unraveling of that abstraction is a common feature of their decline and fall. The desperate expedients now being pursued to expand the American money supply in a rapidly contracting economy have exact equivalents in, say, the equally desperate measures taken by the Roman Empire in its last years to expand its own money supply by debasing its coinage. The Roman economy achieved very high levels of complexity and an international reach; its moneylenders — we would call them financiers today — were a major economic force, and credit played a sizeable role in everyday economic life. In the decline and fall of the empire, all this went away. The farmers who pastured their sheep in the ruins of Rome's forum during the Dark Ages, as we've seen, lived in an economy of barter and feudal custom, in which coins were rare items more often used as jewelry than as a medium of exchange.

A similar trajectory almost certainly waits in the future of our own economic system, though what use the shepherds who pasture their flocks on the Mall in the ruins of a future Washington, DC will find for vast stacks of Treasury bills is by no means clear. How the trajectory will unfold is anyone's guess, but the possibility that we may see sharp declines in the value of the dollar, and of dollar-denominated paper assets, probably should not be ignored, and cashing in abstract representations of wealth for things of more enduring value might well belong high on the list of sensible preparations for the future.

The End of Investment

Many people nowadays, of course, assume that collecting large amounts of money is the best preparation for a difficult future. The thought experiment of the economists on the desert island, though, points to the core problem with that easy assumption. Grant for a moment that the island has a water supply and enough natural foodstuffs that the economists don't have to worry about starving to death. Will the economists on the island have a standard of living corresponding to the one million dollars they each have with them? Of course not; their prosperity will be measured by the breadfruit they pick, the fish they catch, the huts they make and so on.

Once again, money is not wealth. It is a measure of wealth, and it also functions as a social mechanism for distributing wealth. It means nothing unless there is real wealth—actual, nonfinancial goods and services produced by the primary and secondary economies—to back it up. In a healthy market economy, there's a rough balance between the amount of money in circulation and the amount of real wealth produced annually, and so the confusion between money and wealth can slip by unnoticed. When money and wealth get out of sync with one another, problems sprout.

The economic history of the nineteenth century offers a good example. The rising industrial economy of the time drove a massive increase in the production of real wealth. Most industrial nations, though, inherited money systems backed by gold reserves that offered few options for expanding the money supply to match the supply of real wealth. The result was a deflationary spiral that brought major economic depressions every couple of decades for most of the century. In response, in the twentieth century, nation after nation abandoned the gold standard's straitjacket and retooled their money systems to meet the needs of an expanding economy.

That's the context of the present crisis because, in terms of real wealth, we no longer have an expanding economy. The production of real wealth in the world's industrial nations has been in decline now for decades. Some of the deficit has been made up by importing real wealth from overseas, but not all; compare the lifestyle available to a single-salary working-class American family in 1969 to the lifestyle available to a similar family today and it's possible to get a glimpse of just how much stealth impoverishment has taken place over the last 40 years.

This impoverishment went unnoticed by most people because the money supply didn't follow suit. Until the economy came unglued in the second half of 2008, money—most of it, to be sure, in the form of unearned credit and unsecured debt—had never been so abundant or readily available. Some of it got spent on real wealth, which is why real estate and other commodities soared to giddy heights, but most of it was diverted instead into various forms of abstract tertiary wealth. These have become the basis for the most vaporous dimension of today's economic life, the world of investment.

The confusion between money and wealth and the biases imposed by the long economic expansion of industrialism have made it almost impossible to talk sensibly about investments these days. To most people nowadays, it seems perfectly reasonable to think that under normal circumstances, a dollar invested today should always yield more than a dollar tomorrow. In an expanding economy, this is true more often than not, and since people in the industrial world have lived in expanding economies since time out of mind, it seems like a law of Nature that money ought to make money.

The long economic expansion of the industrial age has thus fostered the massive expansion of what old-fashioned Marxists used to call the rentier class — a class whose money makes money for them. Even among people who work for a living, the idea of joining the rentier class on retirement, and living comfortably off investments, has become very popular in recent years. The problem, of course, is that the age of industrial expansion is over. It was made possible in the first place only by exponentially increasing the use of fossil fuels and other natural resources. Like all exponential growth curves, it faced an inevitable collision with the limits of its environment — and that collision is happening around us right now.

We are thus entering a period of prolonged economic contraction — not a recession, or even a depression, but a change in the fundamental dynamic of the economy. Over the centuries just past, a rising tide of economic growth was interrupted by occasional periods of contraction; over the centuries ahead, the long decline of the industrial economy will doubtless be interrupted by occasional periods of relative prosperity. Just as a rising tide lifts all boats, a falling tide lowers them all, and if the tide goes out far enough, a great many boats will end up high and dry. The result is that money can no longer be counted on to make money, and as this becomes apparent, the basis for most of today's rentier class will go out of existence.

Members of the rentier class, whether they earn all their living or only part of it from investments, have made desperate attempts to avoid dealing with this unwelcome reality. These attempts, however, have had the ironic result of making the situation much worse than it had to be. As actual investments in productive economic activities stopped yielding a noticeable profit, more and more investors sought to make money by means of speculation, calling into being a menagerie of exotic financial livestock notable for its complete disconnection from the economy of goods and services. The result was a series of classic speculative bubbles, culminating in the crash of 2008 and the crisis still unfolding around us as I write this. In the process, eager investors who might have lost their money slowly over a period of years have, instead, lost it all at once.

At this point the twilight of investment is very nearly upon us. Most of the assets that attract the attention of investors and speculators alike—from stocks, bonds and mutual funds right up to the most exotic class of derivatives—are part of the torrent of unpayable debt presently flooding the global economy, and the rest are priced at levels that assume that most if not all of that unpayable debt can still be cashed in for goods and services. One way or another, those assets will sooner or later move toward their real value, which in the case of speculative assets is nothing, and in the case of productive assets is much less than they're worth on paper right now. This means that on average, investments will tend to lose money for the foreseeable future.

Interestingly, this is likely to be true even of vital commodities such as crude oil. 2008's price spikes in oil and other energy resources were only partly a product of geological limits. The soaring demand of an overheating economy and speculative money flooding into any asset that was gaining in price both played major parts. Prices collapsed when the speculative money flowed back out, and slumping demand has helped keep prices well below their 2008 peaks since then. As the economy unravels further, the chance of further declines can't be dismissed. It has too rarely been noticed that there are at least two ways to price oil out of the market: the first is for the price per barrel to soar out of reach, while the second is for the economy to contract so sharply that even a low price per barrel is more than most people can pay.

The twilight of investment is one example of the way that metaphysical abstractions woven into the way we think about money can have profoundly concrete impacts on everyday life. A far more important example, however, can be traced in the mismatch between contemporary economics and the realities of energy. Only by grasping the difference between how energy works, and how economists think it works, is it possible to grasp the shape of the predicament of industrial society in any meaningful way, and begin to see how that predicament might be faced.

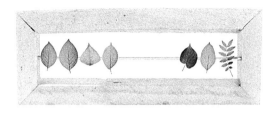

THE PRICE OF ENERGY

THE MISMATCH BETWEEN the tertiary economy of money and the primary and secondary economies of goods and services impacts every dimension of our economic life at the end of the age of cheap abundant energy. Still, the dimension that will have the most direct impact on most of us is that of a collapsing money economy on energy itself. The potential for trouble here can be gauged from a glance back over the way the peak of global oil production has already affected the price of energy.

In the run-up to the 2005 peak of world conventional petroleum production, nearly everyone who was concerned about the world's energy future agreed that the law of supply and demand would define the world's response to the end of cheap oil. As existing reserves depleted, the idea went, the collision between rising demand and shrinking supply would drive prices up. Some people insisted that this would lead to more drilling, more secondary extraction and other measures that would produce more oil and bring the price back down; others insisted that this would lead to the discovery of new energy resources of one kind or another, which would produce more energy and bring the price back down;

still others insisted that this would finally make renewable energy cost-effective, and at least keep the price from rising further; and a minority of pessimists argued that none of these things would happen and the price of oil would rise steadily on up into the stratosphere.

None of them were right. Instead, starting a few years before the peak, oil prices began to move up and down in jolts of increasing violence, culminating in 2008 with a drastic price spike driven by speculative greed, followed by an equally drastic crash driven by speculative panic. The 2008 spike did drive some additional production of oil, mostly from old wells that were uneconomical when oil was $30 a barrel but made plenty of economic sense when it topped $130 a barrel, but most of that additional production was shut back down during the crash that followed.

Meanwhile, the economic shock waves from petroleum's boom and bust played an important role in kicking the props out from under an already shaky global economy, making capital for new energy ventures hard to raise and forcing governments to cut back on the subsidies that have helped make alternative energy technologies economically viable. The results of that impact are still spreading through the world's economies as I write this. One thing that's already become clear amid the bouncing rubble, though, is that models of the future that assumed a steady upward rise in prices have failed to predict the much more complex reality of spike and crash that is shaping our energy future.

Somewhere in the American midwest, perhaps, where a half-completed ethanol plant whose parent company has gone bankrupt is being sold for scrap, and oil leases bought for sky-high prices in 2008 sit unused because the current price of oil won't pay for their development, the dream of a smooth market-driven transition to a different energy system is rolling across a field with the tumbleweeds. Meanwhile the price of oil is continuing its stubborn refusal to obey the law of supply and demand. Demand has dropped, as consumers and businesses caught in the economic downdraft cut consumption, and stockpiles are ample, but the price of oil has more than doubled since its 2009 low, following a slow, ragged, but unmistakable upward trend.

Oil has shown the same habit of standing economic rules on their heads before. Back in the 1970s, one of the great challenges facing the economics profession was the riddle of stagflation. According to one of the most widely accepted rules of macroeconomics, inflation and deflation — which can be defined as the expansion and contraction, respectively, of the money supply relative to the supply of goods and services — form two ends of a continuum of economic behavior. Rising prices, rising wages, and increased economic activity leading to overproduction are all signs of inflation, while flat or declining prices and wages and diminished economic activity leading to recession are all signs of deflation. In the wake of the seventies oil shocks, though, the industrial world found itself in the theoretically impossible situation of an inflationary recession: prices were rising but wages struggled to keep pace and economic activity declined sharply.

That was stagflation. For more than a decade, economists tried to make sense of the riddle it posed, before finally giving up with a certain amount of relief once the Reagan-Thatcher years arrived, deciding that it was an anomaly that had gone away and so didn't matter any more. To many economists who tried to make sense of stagflation, it was clear that the oil crises had something to do with it, but this in itself posed its own awkward questions. The economics of commodity prices have been studied exhaustively since the time of Adam Smith, but the behavior of the world economy in the face of rising oil prices violated everything economists thought they knew.

Only a few economists at the time, and even fewer since, realized that these perplexities pointed to weaknesses in the most basic assumptions of economics itself. E. F. Schumacher was one of these. He pointed out that for a modern industrial society, energy resources are not simply one set of commodities among many others. They are the ur-commodities, the core resources that make economic activity possible at all, and the rules that govern the behavior of other commodities cannot be applied to energy resources in a simplistic fashion. Commented Schumacher:

> I have already alluded to the energy problem in some of the other chapters. It is impossible to get away from it. It is impossible to overemphasize its centrality. [...] As long as there is enough primary energy — at tolerable prices — there is no reason to believe that bottlenecks in any other primary materials cannot be either broken or circumvented. On the other hand, a shortage of primary energy would mean that the demand for most other primary products would be so curtailed that a question of shortage with regard to them would be unlikely to arise.[1]

If Schumacher is right — and events certainly seem to be pointing that way — another of the basic flaws of contemporary economic thought comes into sight. The attempt to treat energy resources as ordinary commodities misses the crucial point that *energy follows laws of its own* that are distinct from the rules governing economic activities and trump these latter rules. Trying to predict the economics of energy without paying attention to the laws that govern energy itself — the laws of thermodynamics — thus consistently yields high-grade nonsense.

One implication that has to be grasped is that the rules that apply to other natural resources, such as minerals, do not work when applied to energy. As North America's deposits of high-grade iron ore became exhausted, the iron industry simply switched over, as already mentioned, to lower grades of ore; these contain less iron per ton than the high-grade ores but are much more abundant, and improved technology for extracting the iron makes up the difference. In theory, at least, the supply of iron ore can never run out, since industry can simply keep on retooling to use ever more abundant supplies of ever lower-grade ores, right down to the modest traces of iron salts dissolved in the sea.

Try to do the same thing with energy, by contrast, and you will not get very far. The only way the iron industry can use lower grades of ore is by using larger amounts of energy per ton of iron produced, and the same rule applies across the board: the lower the concentration of the resource in its natural form, the more energy has to be used to extract it and turn it into useful forms. If you try to do this with energy, you very quickly reach the point at which the energy needed to extract and process the resource is greater than the energy you get from it. Once this point arrives, the resource stops being a source of energy; you might as well try to support yourself by buying $1 bills for $2 each.

Where energy is concerned, concentration counts for much more than quantity. That difference is a consequence of the second law of thermodynamics, which states that the energy in a whole system always moves from more concentrated forms to more dispersed forms, doing work on the way, and the higher the difference in concentration, the more work can be done. In subsets of the system, energy can move against the flow of entropy, from low concentrations to higher ones, but the process always involves more energy moving from higher to lower concentration somewhere else in the same system. The energy concentrated in fossil fuels got there by precisely this process: hundreds of millions of years of sunlight falling on prehistoric plants flowed through the biosphere and was dispersed as waste heat so that a small fraction of that energy could be concentrated in a few locations in the biosphere and buried underground in the form of fossil carbon.

This is why those who expect the universe to hand us some new and improved energy source to replace our dwindling reserves of fossil fuels are fooling themselves. It took an extraordinarily complex series of processes, more time than the human mind has evolved the ability to grasp, and an equally unimaginable amount of energy input from sunlight and the Earth's gravity and internal heat, to produce the highly concentrated fossil fuels we've wasted so casually over the last 300 years. At the beginning of the industrial age, those fuels were the most highly concentrated naturally occurring energy sources that existed on any scale anywhere in this end of the known universe. Even in their current state of depletion, they still are.

Now it's true that there are plenty of other energy sources left as fossil fuels deplete, but all of them are either much scarcer or much more diffuse than the fossil fuels. This is as true of fissionable uranium, which occurs in little more than trace amounts in raw uranium ore, as it is of renewable sources such as sun and wind. Any of these can be concentrated until they will power modern technology, which is designed to use highly concentrated energy sources. Still, there is a catch: you need to use up a great deal of concentrated energy to concentrate the diffuse sources into a form that can be used in our technologies, and once you subtract the costs of concentration from the energy supplies available to a post-fossil fuel world, there generally isn't much left.

Energy Follows Its Bliss

At this point it's necessary to take a closer look at the least popular of all the laws of physics, the second law of thermodynamics, more popularly known as the law of entropy. To call this law unpopular is not to say that it suffers from any lack of recognition from scientists. A comment by Sir Arthur Eddington, one of the twentieth century's greatest physicists, is typical: "If your theory is found to be against the second law of thermodynamics, there is nothing for it but to collapse in deepest humiliation"[2] — a summing-up so useful that it probably deserves to be called Eddington's Law. Entropy is the gold standard of physics, the one thing you can count on even when the rest of the cosmos seems to be going haywire. What makes it unpopular is simply that it stands in stark conflict with some of the most deeply and passionately held convictions of modern industrial humanity.

For all that, it's a simple concept to grasp. Every time you pour yourself a cup of hot coffee on a cold morning you experience entropy in action. Over the following 30 minutes or so, the coffee will get colder and the air around it will get very slightly warmer, until both are the same temperature. Left to itself, energy always moves from higher to lower concentrations: that's the second law of thermodynamics. As it makes its way from higher to lower, the energy can be made to do useful work, and if you're clever enough you can even force some energy up to a higher concentration by allowing a larger amount of energy to go to a lower one, but entropy's price must always be paid.

Human beings don't like to think in these terms, and for the last 300 years, most people in the industrial world haven't had to. Beginning with the eighteenth-century breakthroughs that allowed coal to be turned into steam power, fossil fuels and the technologies built to use them have given human beings command over amounts and concentrations of energy never before wielded by our species, and convinced most people in the industrial world that energy was basically free for the taking. In the halcyon days of industrialism, it was all too easy to forget that what looked like a vast abundance of energy was a cosmic rarity, a minor and finite backwash in one corner of a solar system where energies flowed on a scale too great for human beings to comprehend.

Most people in the industrial world thus find it next to impossible to think of energy as a limited commodity. The resulting assumptions run deep in our culture. Most bright American ten-year-olds, for example, as soon as they learn about electric motors and generators come up with the scheme of hooking up a motor and a generator to the same axle, running the electricity from the

generator back to the motor, and using the result to power a vehicle. It seems logical: the motor drives the generator, the generator powers the motor, perpetual motion results, you hook it up to
wheels, and away you drive on free energy. I was one of those ten-
year-olds, and one of the boxes in my basement may still contain
the drawings I made of the car I planned to build when I turned 16,
using that technology for the engine.

Of course it didn't work. Not only couldn't I get the device
to power my bicycle — that was the test bed I planned to use for
my invention — I couldn't even make the thing run with no load
connected to it at all. No matter how carefully I hooked up a toy
motor to a generator salvaged from an old bicycle light, fitted a
flywheel to one end of the shaft and gave it a spin, the thing turned
over a few times and then slowed to a halt. What annoyed me at
the time and fascinates me now is that the adults I told about my
project understood that it wouldn't work even without a load, and
told me so, but had the dickens of a time explaining why it couldn't
work in terms that a bright ten-year-old could grasp.

This isn't because the subject is overly complicated. The reason
why perpetual motion won't work is breathtakingly simple; the
problem is that the way most people nowadays think about energy
makes it almost impossible to grasp the logic involved. We all learn
in school that energy can neither be created or destroyed, and we
also learn that energy can be defined as the capacity to do work.
Most people, even when they don't think their way through the
logic involved, combine those two statements and end up thinking
that the capacity to do work can neither be created nor destroyed,
and so there's always enough energy on hand to do whatever work
you happen to want done. The only problem with this sanguine
view is that the real world doesn't work this way.

The reason why the real world doesn't work this way is the
second law of thermodynamics. Because of that law, how much
energy you happen to have on hand doesn't tell you how much

work you can get out of it. The amount of work you get out of an energy source depends, not on the amount of energy contained by the source, but on the difference in energy concentration between the energy source and the environment.

Please read that again: *The amount of work you get out of an energy source depends, not on the amount of energy contained by the source, but on the difference in energy concentration between the energy source and the environment.*

Got that? Now let's take a closer look at it.

Left to itself, energy always moves from more concentrated states to less concentrated states; this is why the coffee you poured a few pages back gets cold if you leave it on the table too long. The heat that was in the coffee still exists, because energy is neither created nor destroyed; on the other hand, it's become useless to you, because most of it's dispersed into the environment, raising the air temperature in your dining room by a fraction of a degree. There's still plenty of heat in the coffee, since it stops losing heat when it reaches room temperature and doesn't continue down to absolute zero, but room temperature coffee is not going to do the work of warming your insides on a cold winter morning.

In a very small way, as you sit there considering your cold coffee, you're facing an energy crisis. The energy resources you have on hand (the remaining heat in the coffee) will not do the work you want them to do (warming your insides). Notice, though, that you're not suffering from an energy *shortage*—there's exactly the same amount of energy in the dining room as there was when the coffee was fresh from the coffeepot. No, what you have is a shortage of the difference between energy concentrations that will allow the energy to do useful work.

How do you solve your energy crisis? One way or another, you have to increase the energy concentration in your energy source relative to the room temperature environment. You might do that by dumping your cold coffee down the drain and pouring yourself a fresh cup, say, or by putting your existing cup on a cup warmer. Either way, though, you have to bring a further supply of concentrated energy to bear on the task.

Any time you make energy do anything, some of it will follow its bliss and pass from a higher concentration to a lower one. You can use this effect by allowing a smaller amount of highly concentrated energy to disperse, or by allowing a much larger amount of modestly concentrated energy to do so, or anything in between. To continue the coffee example, you can warm your insides with a small amount of piping hot coffee or a larger amount that has cooled down somewhat. One way or another, though, the total difference in energy concentration between source and environment decreases when work is done. Mind you, you can make energy do plenty of tricks if you're willing to pay its price; no matter what you do, though, you end up with less concentrated energy than you had when you started.

This is why my great discovery at age ten didn't revolutionize the world and make me rich and famous, as I briefly thought it would. Electric motors and generators are ways of turning energy from one form into another — from electricity into rotary motion, on the one hand, and from rotary motion into electricity on the other. Each of them necessarily disperses some energy into waste heat in the process. Thus the amount of electricity that you get out of the generator when the shaft is turning at any given speed will always be less than the amount of electricity the motor needs to get the shaft up to that speed.

This is what gets missed when people assume that the amount of energy, rather than its concentration, is what matters. The latent heat in the waters of the world's oceans, for example, could theoretically provide enough power for the world's economy to keep it running for some preposterously long period of time if it were possible to extract and use it, and every so often somebody tries to build a machine to run on that energy. All these machines fail, because it takes energy to concentrate energy, and the energy you need to concentrate diffuse heat into a form where it can do something useful is more than the energy you get out of the process.

One common way to avoid thinking about our energy predicament, that follows along very similar lines, is to cite the quantity of energy that arrives on Earth by way of sunlight every day and note that it's vastly greater than the quantity of energy our civilization uses in a year. That's true enough, but it misses an essential point: by the time sunlight crosses 93 million miles of space to get to us, it has become dispersed enough that it can do only modest amounts of work in its current form, and concentrating it is very costly in terms of energy. This is why it takes millions of dollars and complex motorized mirror systems to focus sunlight onto a boiler and produce steam, when this could be produced with much simpler technology and at much lower cost by burning fossil fuels: the energy in the fossil fuels is more concentrated, and thus much less energy is needed to put it to work.

More generally, any time you suggest in public that energy forms a crucial limiting factor for civilization, you can safely assume that somebody will respond by insisting that the amount of energy in the universe is infinite. Now of course Garrett Hardin was quite right to point out in *Filters Against Folly* that when somebody says "X is infinite," what's actually being said is "I refuse to think about X." The word "infinite" functions as a thought-stopper, a way to avoid paying attention to something that's too uncomfortable to consider closely.[3]

Still, claiming that there's infinite energy in the universe misses the point in another way as well. Whether or not there is an infinite amount of energy in the universe — and we simply don't know one way or the other — we can be absolutely sure that the amount of highly concentrated energy in the small corner of the universe we can access is sharply and distressingly finite. Since energy always tries to follow its bliss, highly concentrated energy sources are very rare, and only occur when very particular conditions happen to be met.

In the part of the cosmos that affects us directly, one set of those conditions exists in the heart of the sun, where gravitational pressure squeezes hydrogen nuclei so hard that some of them fuse into helium. Another set exists here on the Earth's surface, where plants concentrate energy in their tissues by tapping into the flow of energy dispersing from the sun and other living things do the same thing by tapping into the energy supplies created by the plants. Now and again in the long history of life on Earth, a special set of conditions have allowed energy stockpiled by plants to be buried and concentrated further by slow geological processes, yielding the fossil fuels we use so recklessly. Kinetic energy from water and wind are also powered by sunlight, though the energy they contain is in much less concentrated form. Radioactive elements, which produce geothermal heat and can also be harnessed directly in nuclear power, provide a certain amount of nonsolar energy, but the vast majority of the energy we have on hand here on Earth comes directly or indirectly from the sun.

That in itself defines our problem neatly. As already mentioned, by the time it gets through 93 million miles of deep space, then filters its way down through the Earth's relatively murky atmosphere, the energy in sunlight is pretty thoroughly dispersed; each individual photon has the same energy it had when it left the sun's atmosphere, to be sure, but the photons are much more spread out than they were inside the orbit of Mercury. The low concentration of solar energy reaching Earth is among the reasons why green plants stockpile less than one percent of the energy in the light striking their leaves. Sunlight just isn't that concentrated, and if you want it to do work, you have to invest a great deal of energy to concentrate it first.

A Crisis of Concentration

The importance of concentration in determining how much work a given energy source can do is basic thermodynamics, of the sort that every American high school student used to learn in physics class back in those long-departed days when American high schools had physics classes worth the name. Still, the habits of thinking made automatic by so many years of abundant fossil fuels have made the implications of that principle difficult to grasp. To people in the industrial world, at least, the results of applying the principle of concentration are often counterintuitive; plenty of people miss them, and fumble predictions as a result.

The current brouhaha over anthropogenic climate change offers a good example. A great deal of concentrated fertilizer has been heaped over the issues by propaganda factories on all sides of that debate, but beneath it all is the tolerably well documented fact that we're in the middle of a significant shift in global climate, focused on the north polar region of the globe. The causes of that shift are by no means entirely settled, but it seems a little silly to insist, as some people do, that the mass dumping of greenhouse gases into the atmosphere by humanity can't have anything to do with it — or, for that matter, that it's a good idea to keep on dumping those gases into an atmospheric system that may already be dangerously unstable for reasons of its own.

Still, for the next decade or more, that bad idea is very likely to remain standard practice around the world, and one reason for that is that climate change activists became their own worst enemies once they defined the crisis as "global warming." That seems sensible enough — after all, we're talking about an increase in the total quantity of heat in the Earth's atmosphere — but here as elsewhere, the fixation on quantity misses the crucial point at issue.

Thomas Friedman scored a bull's-eye in his book *Hot, Flat, and Crowded* when he pointed out that what we're facing isn't global warming but "global weirding": not a simple increase in

temperature, but an increase in all kinds of unexpected and dis-
ruptive weather events.[4] As the atmosphere heats up, the most
important effect of that shift isn't the raw increase in temperature;
rather, it's the increase in the difference in energy concentration
between the atmosphere and the oceans. The thermal properties
of water make the seas warm up much more slowly than the air
and the Earth's land surface. Thus even a fairly modest change in
the quantity of heat causes a much more significant change in the
capacity of the atmosphere to do work, since what determines energy's capacity to do work, again, is the difference in concentration
between the energy source and its environment.

The work that the Earth's atmosphere does is called "weather."
From a physicist's point of view, the atmosphere is a titanic engine,
using the temperature difference between sunlight and deep space
to pump huge masses of air and water vapor around the planet.
The better the insulation on the engine — and from a physics perspective the greenhouse effect is simply a way of improving the
insulation on the atmosphere — the more efficiently it can concentrate heat and moisture in some places and take them away from
others. Thus the most visible result of a relatively rapid rise in the
heat concentration of the atmosphere isn't a generalized warming.
Rather, it's an increase in extreme weather conditions on *both* ends
of the temperature scale.

This isn't a new point. It has been made repeatedly by some climate scientists and, interestingly enough, by large insurance companies as well. The huge reinsurance firm Munich Re, for example,
pointed out a few years back that at the current rate of increase, the
annual cost of natural disasters caused by global climate change
would exceed the gross domestic product of the world well before
the end of the twenty-first century.[5] Had climate activists taken
that as their central theme, the abnormally harsh storms in the
eastern half of North America in the winter of 2009–2010 would
have provided plenty of grist for their mills; even hardcore skep-

tics, as they shoveled snow from their driveways for the fourth or fifth time in a row, might have started to wonder if there was something to the claim that greenhouse-gas dumping was causing the weather to go wild. Instead, seduced by our culture's fixation on quantity, most climate advocates defined the problem purely as a future of too much heat, and those same skeptics, shoveling those same driveways, rolled their eyes and wished that a little global warming would show up to help them out.

It may be too late for climate change activists to switch their talking points from global warming to global weirding and be believed by anybody who isn't already convinced, and so we'll likely have to wait until the first major global climate disaster before any significant steps get taken. Still, the same confusion between energy quantity and concentration pervades nearly all of today's discussions about renewable energy. It's easy to insist, for example, that the quantity of solar energy falling annually on some fairly small fraction of the state of Nevada, let's say, is equal to the quantity of energy that the US uses as electricity each year, and to jump from there to insist that if we just cover a hundred square miles of Nevada with mirrors so all that sunlight can be used to generate steam, we'll be fine.

What gets misplaced in appealing fantasies of this sort? Broadly speaking, three things.

The first is the usual nemesis of renewable energy schemes, the problem of net energy. It would take a substantial amount of highly concentrated energy to build that hundred-square-mile array of mirrors, counting the energy needed to manufacture the mirrors, the tracking assemblies, the pipes, the steam turbines and all the other hardware, as well as the energy needed to produce all the raw materials that go into them. It would take another very large amount of concentrated energy, regularly supplied, to keep it in good working order amid the dust, sandstorms and extreme temperatures of the Nevada desert. If the amount of energy produced by the scheme came anywhere close to what's theoretically possible, this would probably be the only time in history that this has ever occurred with a very new, very large, and very experimental technology. Subtract the energy cost of building and running the plant from the energy you could reasonably (as opposed to theoretically) expect to get out of it, and the results will inevitably be a good deal less impressive than they look when presented on paper by enthusiasts.

The second is another equally common nemesis of renewable energy schemes: the economic dimension. Plenty of renewables advocates say, in effect, that people want electricity, and a hundred square miles of mirrors in Nevada will provide it, so what are we waiting for? This sort of thinking is extremely common, of course. Mention that any popular technology you care to name might not be economically viable in a future of energy and resource constraints, and you're sure to hear plenty of arguments that it has to be economically feasible because, basically, it's so nifty. There's a reason for this blindness to the economic dimension. It's a natural human response to the relative ease of accessing capital, materials and labor in an age of abundance, like the one that's coming to an end around us right now.

The end of that age, though, makes this way of thinking a hopeless anachronism. In an era of energy and resource constraints, any proposed use of energy and resources must compete against all other existing and potential uses for a supply that is no longer adequate to meet them all. Market forces and political decisions both play a part in the resulting process of triage. If investing billions of dollars (and, more importantly, the equivalent amounts of energy and resources) in mirrors in the Nevada desert doesn't produce as high an economic return as other uses of the same money, energy and resources, the mirrors are going to draw the short straw. Political decisions can override that calculus to some extent, but impose an equivalent requirement: if investing that money, energy and resources in mirrors doesn't produce as high a political payoff as other uses of the same things, once again, the fact that a hundred square miles of mirrors might theoretically allow America's middle classes to maintain some semblance of their current lifestyle is not going to matter two photons in a Nevada sandstorm.

Barbarism and Good Brandy

Still, the failure to grasp the implications of net energy, on the one hand, and economic triage on the other, both unfold from a deeper failure to think through the implications of thermodynamic reality. These implications comprise the third thing that's consistently misplaced in fantasies of the mirrors-in-Nevada variety. The solar energy that reaches the Earth is relatively diffuse. It takes a great deal of hardware to concentrate it to the point that you can run today's technologies with it, and that hardware has to be paid for in terms of energy as well as economics. This is not a new problem, either: big solar power schemes, of the sort now being proposed, were repeatedly tried from the late nineteenth century on, and just as repeatedly turned out to be economic duds.

Consider the solar engine devised and marketed by American engineer Frank Shuman in the first decades of the twentieth century. The best solar engine of the time, and still the basis of a good many solar power designs today, it was an extremely efficient device that focused sunlight via parabolic troughs onto water-filled pipes that drove an innovative low-pressure steam engine. Shuman's trial project in Meadi, Egypt, used five parabolic troughs 204 feet long and 13 feet wide. The energy produced by this sizable and expensive array? All of 55 horsepower.[6] Modern technology could do better, doubtless, but not much better, given the law of diminishing returns that affects all movements in the direction of efficiency, and the evidence of solar power schemes to date suggests that the improvements will not be enough to matter.

Does this mean that solar energy is useless? Not at all. What it means is that a relatively diffuse source of energy, such as sunlight, is not an efficient way to run the kind of technology created to use a relatively concentrated source such as coal. That's what Shuman was trying to do. Like most of the solar pioneers of his time, he had done the math, realized that fossil fuels would run out in the not infinitely distant future and argued that they would have to be replaced by solar energy. "One thing I feel sure of," he wrote, "and that is that the human race must finally utilize direct sun power or revert to barbarism."[7]

He may well have been right, but trying to make lukewarm sunlight serve the same functions as the blazing heat of burning coal is not an effective response. The difficulty, as already explained, is that whenever you turn energy from one form into another, you inevitably lose energy to waste heat in the process, your energy concentration goes down accordingly and so does the amount of work your energy can do. If you have abundant supplies of a highly concentrated fuel such as coal or petroleum that doesn't matter; but with a diffuse energy source such as sunlight, it impacts the economic feasibility of the project very quickly. This is why Shuman's solar plant, which covered well over 13,000 square feet, produced less power than a very modest diesel engine that cost a small fraction of Shuman's costs to build and operate.

Now of course there have been significant improvements on the efficiency of some of the processes involved in those early solar engines. Still, a good many of the basic limits that the nineteenth and early twentieth century solar pioneers faced are not subject to technological improvement because they unfold from the difference between diffuse and concentrated energy.

Nearly 30 years ago, when I was studying appropriate technology in college, a standard example of thermodynamic limits was ordinary geothermal heat. This is the gentle warmth that filters up through the Earth's crust from the mantle many miles below, where there isn't any underground magma close enough to the surface to set off geysers and make commercial geothermal plants an option. There is a huge amount of it, but it's very diffuse, and as a result a few simple calculations will show that the energy you get from pumping the heat to the surface and using it to drive an engine will be less than the energy needed to run the pumps. On the other hand, if all you want is diffuse heat, you're looking in the right place; hooking up a heat pump to a hole in the ground and using the results for domestic heating and cooling has proven to be an efficient and economical technology in recent years.

The same thing is true for OTEC, another of those ideas whose time is always about to come and never quite arrives. The acronym stands for Oceanic Thermal Energy Conversion, and it is supposed to do with the thermal difference between deep and surface water what a geothermal power plant does with the thermal difference between hot rocks half a mile down and the cold surface of the planet. You can run a heat engine on OTEC power, but it takes about two thirds of the power the plant can generate to run the pumps that bring cold water up from the depths. That means the plant has a net energy of 0.33 or so, even before factoring in the energy cost of the OTEC plant; in practical terms, what it means is that you fund your plant with government grants or you go broke. On the other hand, there's at least one resort in the Pacific that uses OTEC for the far simpler task of air conditioning. Again, if all you need to do is move diffuse heat around, a diffuse energy source is more than adequate; it's when you need to do something more complex that you have problems.

A closer look at why that happens may be useful here. The core concept to grasp is that, for reasons hardwired into the laws of thermodynamics, converting energy from one form into another is highly inefficient, with a great deal of the original energy being lost to diffuse heat. A heat engine is an energy converter; it takes in concentrated heat and puts out mechanical energy through a rotating shaft, which is then hooked up to a drive train, a propeller, a generator or some other application. Of all the energy released by burning gasoline in an average automobile engine, which is a form of heat engine, only around 25 percent goes into turning the crankshaft; the rest is lost as diffuse heat. If you're smart and careful, you can get a heat engine to reach efficiencies above 50 percent; a modern combined-cycle electrical generation plant working at top efficiency can hit 60 percent, but that's about as good as the physics of the process will permit.

Most other ways of turning one form of energy into another are no more efficient than heat engines, and many of them are much less so. (This is why heat engines are used so extensively in modern technology; inefficient as they are, they're better than nearly all of the alternatives.) The reason nobody worries much about these inefficiencies is that our heat engines run on fossil fuels and fossil fuels contain so much energy in so concentrated a form that the inefficiencies aren't a problem. Seventy-five percent of the energy in the gas you pour into your car gets turned into diffuse heat and dumped via the radiator, but you don't have to care; there's still more than enough to keep you zooming down the road.

With alternative energy sources, though, you do have to care. That's why the difference between diffuse and concentrated energies matters so crucially; not only specific technologies, but whole classes of technologies on which the modern industrial world depends embody such massive inefficiencies that diffuse energy sources won't keep them running without losses that make them hopelessly uneconomical. Lose 75 percent of the energy in a gallon of gasoline to waste heat, and you can shrug and pour another gallon in the tank; lose 75 percent of the energy coming out of a solar collector in the same way, and you may well have passed the point at which the solar collector no longer does enough work to be worth the energy and money it cost to build and maintain it.

The one way around this problem is to use the behavior of energy in your favor. Energy turns itself into diffuse heat with very high efficiencies, sometimes approaching 100 percent, and if you set out deliberately to make use of diffuse heat you can take advantage of this fact. This is why using sunlight to heat water, air, food or anything else to temperatures in the low three digits on the Fahrenheit scale is among the most useful things you can do with solar energy and why, when you're starting out with diffuse heat, the most practical things you can do with it usually involve leaving it as diffuse heat.

Augustin Mouchot, the great nineteenth-century pioneer of solar energy, kept running up against this issue in his work.[8] Mouchot's motivations were much the same as Shuman's: he began working with solar energy out of a concern that France, handicapped by its limited reserves of coal, needed some other energy source to compete in the industrial world of the late nineteenth century. He built the world's first successful solar steam engines, but they faced the same problems of concentration that made Shuman's more sophisticated project an economic failure; a typical Mouchot engine, his 1874 Tours demonstration model, used 56 square feet of conical reflector to focus sunlight on a cylindrical boiler, and generated all of one half horsepower.

Yet some of his other solar projects were much more successful. For many years, the French Foreign Legion relied on one of his inventions in their campaigns: a collapsible solar oven that could be packed into a box 20 inches square. It had the same general design as the engine, a conical reflector focusing sunlight onto a cylinder that pointed toward the sun, but it worked, and worked well; the Mouchot oven could cook a large pot roast from raw to well done in under half an hour. Cooking a roast is a process that uses diffuse heat. In most contemporary kitchens we take highly concentrated energy, in the form of electricity or natural gas, and turn that into diffuse heat in an oven, but if what you have is the diffuse heat of sunlight, you can still cook a roast with great efficiency.

Another of Mouchot's inventions, a solar still designed for France's brandy industry, proved equally successful, converting wine into brandy at a rate of five gallons an hour. Interestingly, the still's reliance on diffuse solar heat turned out to have an unexpected benefit. As every moonshiner knows, one common problem in distilling alcoholic beverages is the unpredictable appearance of "hot spots" in the still that give the result a burnt taste; the steady solar heat of Mouchot's still did not have that problem. The still turned out a superior brandy, "bold and agreeable to the taste," Mouchot wrote proudly, "and with...the savor and bouquet of an aged *eau-de-vie*."[9] Again, diffuse sunlight doesn't convert well to mechanical motion via a steam engine, due to the inevitable conversion losses, but it's very efficient as a source of heat.

The point to take away from these examples is that the technology that's useful, appropriate and economically viable in a setting of energy and resource constraints is not the same as the technology that's useful, appropriate and economically viable in a setting of abundance. Centralized power generation is another good example. If you've got ample supplies of highly concentrated energy, it makes all the sense in the world to build centralized power plants to burn fossil fuels, convert the result to electricity and send the electricity across hundreds or thousands of miles to consumers who turn it into diffuse heat. You'll lose plenty of energy at every point along the way, especially in the conversion of one form of energy to another, but if your sources are concentrated and abundant, that doesn't matter much. If concentrated energy sources are scarce and rapidly depleting, on the other hand, this sort of extravagance can no longer be justified, and after a certain point, it can no longer be afforded.

Working With Diffuse Energy

This has very troubling implications for an economic system that has built essentially all its technologies to run on highly concentrated energy. Fossil fuels are as concentrated as they are because the energy they contain was gathered over countless centuries and then concentrated by geological processes involving fantastic amounts of heat and pressure over millions of years. They define the upper end of the curve of energy concentration, at least on this planet. To bring the energy from any other source up to the same concentrations as fossil fuels requires paying entropy's price in a big way; the energy costs of concentration have to be factored in, which is why no other energy resource can compete with fossil fuels in economic terms.

As fossil fuel supplies deplete, an economic system that depends on them for its survival risks finding itself in a very nasty bind. Our industrial economies, certainly, have proceeded to make that bind even worse by using part of the fossil fuels that still remain to concentrate more diffuse resources to the same concentrations as fossil fuels. This is the logic underlying most of the currently popular renewable energy technologies. To make and maintain a solar panel, a big wind turbine or an ethanol plant, it's necessary to use up a great deal of concentrated energy — raw materials must be extracted and refined, a variety of complex manufacturing processes must be performed and the end result is then shipped hundreds or thousands of miles — and all this work has to be done with highly concentrated energy sources, mostly fossil fuels, because that's what our current technology is designed to use.

Thus trying to produce highly concentrated energy out of diffuse sources is a losing bargain that actually increases the drawdown on remaining supplies of concentrated energy. The only other option most economists and mainstream energy mavens have been willing to discuss — ignoring renewable energy sources, continuing to draw down fossil fuels and praying for a scientific miracle to bail us out of the consequences of our own lack of foresight — is even less promising. Still, there is at least one other option, which is to learn to do as much as possible with relatively diffuse energy, while husbanding remaining stocks of highly concentrated energy sources for those necessary tasks that can't be done without them. This offers unexpected possibilities, because many of the things we do with concentrated energy could be done with more diffuse energy instead.

Nearly every home in the industrial world, for example, has hot water on tap. This is by no means a pointless luxury; the contemporary habits of washing dishes, clothes and bodies with ample amounts of hot water and soap have eliminated whole categories of illnesses that plagued our ancestors not that long ago. A very large fraction of those homes get that hot water by burning fossil fuels, either right at the hot water heater or at a power plant that uses the heat to generate the electricity that does the heating. A society that has ample supplies of fossil fuels can afford to do that; a society running out of concentrated energy sources is likely to face increasing troubles doing so.

There's a crucial point not often recognized, however: it doesn't take concentrated energy to heat a tank full of water from ambient temperature to 120° or so. The same thing can be done very effectively by diffuse energy sources such as sunlight.

Enter the solar water heater.

This is arguably the most mature and successful solar technology in existence right now. The process is simple: one of several kinds of collectors gather heat from the sun and transmit it either to water, in places that don't get freezing temperatures, or to an antifreeze solution in places that do. In a water system, the hot water goes from the collector to an insulated tank and eventually to the hot water faucet; in an antifreeze system, the antifreeze circulates through a heat exchanger that passes the heat to water, which then goes into an insulated tank to wait for its moment of glory. In most parts of North America, a well-designed solar hot water system will cut a home's energy use to heat water by 70 percent; in the Sun Belt, it's not at all uncommon for a solar hot water system to render any other hot water heater unnecessary.

It will doubtless occur to my readers that installing a solar hot water system in their homes will not save the world from the consequences of fossil fuel depletion. What it will do is take part of the work now done by highly concentrated energy sources — most of which are rapidly depleting and can be expected to become more expensive in real terms over the decades to come — and hand it over to a readily available energy source of much lower concentration that, among other things, happens to be free. That's an obvious practical gain for the residents that install solar water heating; it's also a collective gain for the community and society, since remaining supplies of fossil fuels can be freed up for more urgent uses or, just possibly, left in the ground where they arguably belong.

Since much of the energy that people use in their daily lives takes the form of diffuse heat — the sort that will heat water, warm a house, cook a meal and so on — it makes more economic sense in an energy-poor society for people to gather that heat right where they are, and put it to work there, rather than using concentrated energy sources from far away. The same point can be made with equal force for a great many industrial processes: when what you need is diffuse heat — and for plenty of economically important activities, such as distilling brandy, that's exactly what you need — sunlight, concentrated to a modest degree by way of reflectors or fluid-heating panels, will do the job quite effectively.

This is the secret to using diffuse energy sources in an economically viable manner: diffuse heat is where energy goes when it follows its bliss, so if you let it do what it's going to do anyway, you can turn a relatively unconcentrated energy source to heat at very high efficiencies. The heat you get is fairly mild compared to burning gasoline, for example, but that's fine for practical purposes. Again, it doesn't take intense heat to raise a bathtub's worth water to 120°, warm a chilly room or cook a meal, and it's precisely tasks like these that solar energy and other diffuse energy sources do reliably and well.

Thus it's curious, to use no stronger word, that so eminently practical a step as installing solar hot water systems has received so little attention to date in the peak oil and climate change communities. It's all the more curious because national governments, which so often seem incapable of encountering a problem without doing their level best to make it worse, have actually done something helpful for a change: there are very substantial incentives in many countries for installing a residential solar hot water system. Why, then, haven't solar hot water heaters blossomed like daisies atop homes across the industrial world? Why haven't activists made a push to define this technology as one part of a meaningful response to the crisis of our time?

It's an interesting question to which there are no definite answers. Partly, I suspect, it ties into the weird disconnection between belief and action that pervades the apocalyptic end of contemporary culture. Of the sizeable number of people in today's America who say they believe that the world is coming to an end in 2012, for example, how many have stopped putting money into their retirement accounts? To judge by the evidence I've been able to gather, not many. In the same way, of the people who say they recognize that today's extravagant habits of energy use are only possible because of a glut of cheap fossil fuels, and will go away as fossil fuels deplete, those who are taking even basic steps to prepare themselves for a future of scarcity and socioeconomic disruption make up an uncomfortably small fraction. It's hard to imagine passengers on a sinking ship glancing over the side to see the water rising and going back to their game of shuffleboard on the deck, but a similar behavior pattern is far from rare these days.

Still, part of the issue may well be the same fixation on quantity discussed already. Solar water heaters don't produce or save a great quantity of energy. Water heating uses around 15 percent of an average North American home's energy bill, and so a solar hot water system that replaces 70 percent of that will account for a bit more than 10 percent of total home energy use. (This is still enough to pay for most professionally installed solar hot water systems in three to seven years of normal use.) If every home in North America put a solar hot water heater on its roof, the impact on energy consumption would be noticeable, but in terms of raw quantity, it wouldn't be huge.

Still, this misses at least three important points. First, of course, installing a solar hot water system can very easily be one piece of a broader program of energy conservation with a much larger impact. Knock 10 percent off household energy use with a solar water heater, another 20 percent by insulating, weatherstripping, and similar efficiency measures, and another 20 percent with lifestyle changes, and your home will be getting by with half the concentrated energy it uses right now. If even a large minority of homes in North America took these steps, or others with similar effects, the effect on energy use would be very substantial indeed.

Second, there's a large and underappreciated difference between essential and nonessential energy uses, and it's one that many of us will learn to recognize in the challenging years ahead. A great deal of energy use in North America today is nonessential—think for a moment of all the energy currently devoted to the tourism industry, a sizable sector of the US economy that could be shut down tomorrow without impacting anything but the unemployment rolls—and a very large amount of that will go away as the world slides down the curve of energy descent. Whether or not hot water is strictly essential, its direct practical benefits in terms of health and comfort put it a good deal closer to the core, and that makes finding ways to provide it more important than, say, finding ways to keep video games powered.

Third, as I've already suggested, we face a shortage of energy concentration rather than energy quantity. That doesn't make our predicament any less severe; given that nearly all modern technology is geared to run on highly concentrated energy, and that the survival of seven billion people depends on the continued operation of the existing technostructure, the rapid depletion of our world's remaining reserves of highly concentrated energy could very easily lead to a grim future even in the presence of vast amounts of diffuse energy. Still, this also means that taking steps to meet as many energy needs as possible using relatively diffuse energy sources can have a disproportionate impact on the way that the future unfolds.

This is where E. F. Schumacher's concept of "intermediate technology" becomes central to a thoughtful response to the predicament of our time. Schumacher's idea, which he applied specifically to the nonindustrial nations of his own time, was that state-of-the-art factories and an economy dependent on exports to the rest of the world are not actually that useful to a relatively poor nation trying to build an economy from the ground up. He was right, of course — those Third World nations that have prospered since his time are precisely the ones that used trade barriers to shelter low-tech domestic industries and entered the export market only after building a domestic industrial base one step at a time — but in a future in which all of us will be a good deal poorer than we are today, his insights have a wider value.

A state-of-the-art factory, after all, is costly in terms much more concrete than those of the tertiary economy; a great deal more concentrated energy has to go into building and maintaining one than you need to build and maintain a workshop that uses hand tools and human muscles to produce the same goods. Thus the concept of intermediate technology, and a great many of the specific technologies Schumacher and his associates and successors created under that banner, provide crucial resources as the age of cheap abundant fossil fuels is drawing to an end. When concentrated energy is scarce, local production of diffuse energy for local use is a far more viable way to support most sectors of the economy.

This will allow what highly concentrated energies are left to be directed to those applications that actually need them, while also shielding local communities from the consequences of the partial or total breakdown of centralized energy systems. The resulting economy may not have much resemblance to today's fantasies of a high-tech future, but the barbarism Frank Shuman feared is not the only alternative to that future. There's something to be said for a society, even a relatively impoverished and resource-scarce one, that can still reliably provide its inhabitants with hot baths, warm rooms in winter and well-done pot roasts — and, as Augustin Mouchot would likely have pointed out, good brandy.

What this means, ultimately, is that the change from today's industrial economy to the economies of the future can't be accomplished by plugging in some other energy source to replace petroleum or other fossil fuels. Nor can it be done by downscaling existing technologies to fit a sparser energy budget. It requires reconceiving our entire approach to technology, starting with the paired recognitions that the very modest supply of concentrated energy sources we can expect to have after the end of the fossil fuel age will have to be reserved for those tasks that still need to be done and can't be done with any more diffuse source, and that anything that can be done with diffuse energy needs to be done with diffuse energy if it's going to be done at all.

A society running on diffuse energy resources will thus not be able to make use of the same kinds of technology as a society running on concentrated energy resources, and attempts to run most existing technologies off diffuse renewable sources are much more likely to be distractions than useful options. In the transition from today's technology dominated by concentrated energy to tomorrow's technology dominated by diffuse heat, in turn, some of the most basic assumptions of contemporary economic thought — and of contemporary life, for that matter — are due to be thrown out the window.

The Economics of Entropy

At this point we can sum up the implications of this chapter in terms of the economic issues central to the argument of this book. The first point that has to be grasped is that the claims made every decade or so that we can power the world off some relatively diffuse energy source need to be added to the same round file as perpetual motion schemes. Today's industrial societies require very highly concentrated energy sources; our transportation networks, our power grids and most of the other ways we use energy all work by dispersing very high concentrations of energy all at once into waste heat, and without the highly concentrated energy resources to power them, those things won't work at all.

The second point is that there are still plenty of economically productive things that can be done with more diffuse energy sources. Solar water heating is a mature and highly successful technology; so is passive solar heating for buildings; so are a good many other uses, such as solar ovens for cooking, water purification and the like. There are many other proven technologies using other forms of diffuse renewable energy, and these are also well worth implementing. All these can contribute mightily to the satisfaction of human needs and wants, but they presuppose social and economic arrangements that have little in common with the centralized energy system of power plants, refineries and power grids we have today. As concentrated energy from fossil fuels becomes scarce, in other words, and diffuse energy from renewable sources has to take up the slack, many of the ground rules shaping today's economic decisions will no longer apply.

This means, in effect, that entropy must find its way into every part of economic thinking. It's worth noting that most of today's economic theories admit entropy, in one way or another, into the secondary economy: it's generally admitted that producing goods and services consumes resources and produces waste and that energy fed into the process is lost to entropy in one way or another. Most current economic thought, however, explicitly rejects the role of entropy in the primary economy, insisting that resources are by definition always available, if you only invest enough labor and capital. As for the tertiary economy, most economic theories accept it as given that tertiary wealth can increase indefinitely over time; it is not merely free from entropy, but moves in the opposite direction.

In the real world, by contrast, the primary economy is just as subject to entropy as the secondary one. Oil that has been pumped out of the ground and burned is no longer available to use as an energy resource, and if enough of it has been pumped out, the oil field runs dry and it stops being a resource too. The Earth can keep some resources available at a steady level by surfing the very slow entropic decline of the Sun, but only if human action doesn't mess up the natural cycles that turn sunlight into renewable resources, and even so those resources are not available in limitless amounts. Ignoring entropy in the primary economy of Nature thus sets conventional economics up for a future of repeated and accelerating failure.

And the tertiary economy? This is where things get interesting, because the claim that tertiary wealth increases over time has been accepted even by most critics of contemporary economics. There has accordingly been a flurry of proposals for changing the way money works so that it loses value over time, or otherwise is reinvented to make it fit the realities of life on a limited and fragile planet. These efforts are understandable, but they are also unnecessary because money as it exists today has an exquisite mechanism for losing value over time. The only difficulty is that mainstream economists and the general public alike treat it with the same shudder of dread and indignation their Victorian ancestors directed toward sex.

We're talking, of course, about inflation.

Inflation is the primary way that the tertiary economy resolves the distortions caused by the mismatch between the limitless expansion of the tertiary economy, on the one hand, and the hard limits ecology and entropy place on the primary and secondary economies, on the other. When the amount of paper wealth in the tertiary economy outstrips the production of actual, nonfinancial goods and services in the other two economies, inflation balances the books by making money lose part of its value. I suspect—though it would take a good econometrician to put this to the test—that in the long run the paper value lost to inflation equals the paper value manufactured by interest on money, once the figures are adjusted for actual increases or decreases in the production of goods and services and for tertiary wealth lost in other ways, such as bond defaults and bank failures.

It's instructive to note, in this context, what happens when governments attempt to stop the natural balancing process of inflation. In American economic history, there are two periods in which this was attempted in a big way; the first was between the Civil War and the First World War, and the second was between 1978 and 2008. In the first of these periods, the US treasury reacted against the rampant inflation of the Civil War years by imposing a strict gold standard on the currency.

Despite the claims of today's precious metal advocates, this did not produce economic stability and prosperity. Quite the contrary, the pace at which new gold entered the economy was less than the rate at which the production of goods and services expanded, and the result was the longest sustained bout of deflation in the history of the United States. The economic terrain of the second half of the nineteenth century was thus a moonscape cratered by disastrous stock market collapses and recurrent depressions. The resulting bank and business failures probably eliminated as much paper value from the economy as inflation would have, but did so in a chaotic and unpredictable way: instead of everybody's corporate bonds losing five percent of their value each year due to inflation, for example, some bonds were paid in full while others became worthless when the companies backing them went out of existence.

The same calculus has come into play since the beginning of the Volcker era at the Federal Reserve Board, when "fighting inflation" became the mantra of the day; since then we've had a succession of crashes as colorful as anything the nineteenth century produced. Thirty years of economic policy dedicated to minimizing inflation have guaranteed a sizable second helping of economic collapse in the years to come. Still, in the longer term, I suspect inflation will also play a major role in the unraveling of the current mess. With the end of the age of cheap abundant fossil fuels, the world faces a very substantial decrease in the amount of primary and secondary wealth in the world, and the paper wealth of the tertiary economy will have to lose value even faster to make up for that decline. Just how this will play out is anyone's guess, but one way or another it's unlikely to be pretty.

What this implies, as we have already seen, is that economic activities never exist in a vacuum. Current economic ideas took shape in an age where economic processes were dominated — one might even say distorted — by extravagant supplies of cheap and highly concentrated fossil fuel energy. The new ground rules of economics that will take shape in the twilight of the age of cheap energy will be shaped, in turn, by the recognition that energy is once again as scarce, costly and diffuse as it has been through most of human history. This recognition may seem obvious enough in the abstract, but in today's economic thought, as we have seen, it is consistently ignored — and never more so than in the theme of the next chapter, the economics of technology.

THE APPROPRIATE TOOLS

PERHAPS THE BEST measure of the ways that three centuries of cheap abundant energy have shaped modern thinking is the forceful resistance so many of us put up to thinking about technology in economic terms. It should be obvious that whether or not a given technology continues to exist in a time of faltering abundance depends on three economic factors. The first is whether the things done by that technology are necessities or luxuries, and if they are necessities, just how necessary they are; the second is whether the same things, or at least the portion of them that must be done, can be done by another technology at a lower cost in scarce resources; the third is how the benefits gained by keeping the technology supplied with the scarce resources it needs measure up to the benefits gained by putting those same resources to other uses. To most people nowadays, though, these points are anything but obvious.

To accept that economic factors might lead societies to discard technologies in an age of contraction flies in the face of some of the most pervasive habits of modern thought. To a remarkable extent, belief in technological progress has taken on a religious value in the modern world, especially — but not only — among those who reject every other kind of religious thinking. The faith in progress has drawn strength from the unquestionable fact that for the last three centuries those who believed in the possibilities of progress have generally been right. There have been some stunning failures, to be sure, but the trajectory that reached its climax with human footprints on the Moon makes a potent argument for the claim that technological complexity is cumulative, irreversible and immune to economic concerns.

Still, it's a mistake to take the experience of an exceptional epoch in human history as a law governing that history as a whole. In the final analysis, the three centuries of exponential growth that put those bootprints on the gray dust of the Sea of Tranquility were made possible by the conjunction of historical accidents and geological laws that allowed a handful of nations to seize the fantastic treasure of fossil carbon buried in the Earth, and burn through it at an extravagant rate, flooding their economies with almost unimaginable amounts of cheap concentrated energy.

It's been fashionable to assume that the arc of progress was what made all that energy available, but there's very good reason to think that this puts the cart in front of the horse. A few technical advances were necessary to break into the Earth's carbon storehouse, but it was the huge surpluses of available energy thus released that made the following age of technological progress possible as inventors, industrialists and ordinary people all discovered that it was cheaper to have machines powered by fossil fuels take over jobs that had been done for millennia by human and animal muscles, fueled by solar energy in the form of food.

The logic of abundance that was made plausible as well as possible by those surpluses has had impacts on our society that very few people have yet begun to confront. Many of the most basic ways that modern industrial societies use energy, for example, make sense only if energy is so cheap and abundant that waste simply isn't something to worry about. Consider the system of electrical generation and distribution that provides most of the energy modern people use outside of transportation. The inefficiencies of electrical generation and distribution are such that less than one-third of the energy contained in the fuel that is burnt in power plants actually reaches the end user; the rest becomes waste heat at various stages in the process. For the sake of having electricity instantly available from sockets on nearly every wall in the industrial world, in other words, we accept unthinkingly a system that requires us to burn more than three times as much energy as we actually use.

In a world where concentrated energy sources are scarce and expensive, many extravagances of this kind will stop being possible, and most of them will stop being economically feasible as well. In a certain sense, this is a good thing, because thinking along these lines suggests many ways in which nations facing an energy crisis can cut their losses and maintain vital systems. It's been pointed out repeatedly, for example, that the electrical grids that supply power to homes and businesses across the industrial world may stop being viable early on in the process of contraction, and some peak oil thinkers have accordingly drawn up nightmare scenarios around the sudden and irreversible collapse of national power grids.[1] Like most doomsday scenarios, though, these rest on the unstated and unexamined assumption that everybody involved will sit on their hands and take no action as the collapse unfolds.

This assumption rests in turn on a very widespread unwillingness to think through the consequences of an age of contracting energy supplies. The managers of a power grid facing collapse due to a shortage of generation capacity have one obvious alternative to hand: cutting nonessential sectors out of the grid, so the load on the grid decreases to a level that available generation capacity can handle. In an emergency, for example, many American suburbs and a large part of the country's nonagricultural rural land could have electrical service shut off completely, and an even larger portion of both could be put on the kind of intermittent electrical service common in the Third World, without catastrophic results. Of course there would be an economic impact, but it would be modest compared to letting the whole grid crash.

Over the longer term, just as the twentieth century was the era of rural electrification, the twenty-first promises to be the era of rural de-electrification. A nation facing prolonged or permanent shortages of electrical generating capacity could make its available power go further by cutting its rural hinterlands off the power grid, and leaving them to generate whatever power they can by local means. Less than a century ago, most farmhouses on North America's Great Plains had a windmill nearby, generating 12 or 24 volts for home use whenever the wind blew; the same approach will be just as viable in the future, not least because windmills on the home scale — unlike the huge turbines central to most current notions of wind power — can be built by hand from readily available materials. (Skeptics take note: I helped build one in college in the early 1980s using, among other things, an old truck alternator and a propeller hand-carved from wood. Yes, it worked.)

Steps like this have seen very little discussion in recent years, because the assumptions about technology that unfold from the faith in progress make them unthinkable. Most people in the industrial world today seem to have lost the ability to imagine a future that doesn't have electricity coming out of a socket in every wall without going to the other extreme and leaning on Hollywood clichés of universal destruction. The idea that some of the most familiar technologies of today may simply become too expensive and inefficient to maintain tomorrow is alien to ways of thought dominated by the logic of abundance.

That blindness, however, comes with a huge price tag. As the age of abundance made possible by fossil fuels comes to its inevitable end, a great many things could be done to cushion the impact.

Quite a few of these could be done by individuals, families and local communities—to continue with the example under discussion, it would not be that hard for people who live in rural areas or suburbs to provide themselves with backup systems, using local renewable energy, to keep their homes viable in the event of a prolonged or permanent electrical outage. None of the steps involved are hugely expensive, most of them have immediate payback in the form of lower energy bills, and local and national governments in much of the industrial world are currently offering financial incentives—some of them very robust—to those who pursue them. Despite this, most of the attention and effort that goes into responses to a future of energy constraints focuses on finding new ways to put power into the existing electrical grid without ever asking whether this will be a viable response to an age when the extravagance of the present day is no longer an option.

This is why attention to economics in a world on the cusp of peak oil is so crucial. Could an electrical grid of the sort we have today, with its centralized power plants and its vast network of wires bringing power to sockets on every wall, remain a feature of life throughout the industrial world in an energy-constrained future? If attempts to make sense of that future assume that this will happen as a matter of course, or start with the unexamined assumption that such a grid is the best (or only) possible way to provide energy to consumers, the resulting discussions will fixate on technical debates about whether and how that can be made to happen, while the core issues that need to be examined slip out of sight. One question that has to be asked instead is whether a power grid of the sort we take for granted will be economically viable in such a future—that is, whether such a grid is as necessary as it seems to us today; whether the benefits of having it will cover the costs of maintaining and operating it; and whether the scarce resources it uses could produce a better return if put to work in some other way.

Local conditions might provide any number of answers to that question. In some countries and regions, where people live close together and renewable energy sources such as hydroelectric power promise a stable supply of electricity for the relatively long term, a national or regional grid of the current type may prove viable. In others, it might be much more viable to have restricted power grids supplying urban areas and critical infrastructure, while rural hinterlands return to locally generated power or to non-electrified lifestyles. In still others, a power grid of any kind might prove to be economically impossible.

Under all these conditions, even the first, it makes sense for governments to encourage citizens and businesses to provide as much of their own energy needs as possible from locally available, diffuse energy sources such as sunlight and wind, and to pursue conservation and efficiency measures much further than they have been pursued to date. Under all these conditions, in turn, it makes at least as much sense for individuals, families and communities to take such steps themselves, so that any interruption in electrical power from the grid — temporary or permanent — becomes an inconvenience rather than a threat to survival.

A case can easily be made that in the face of a future of uncertain energy supplies, off-grid sources of space heating, hot water and other basic necessities are as important in a modern city, suburb or rural area as life jackets are in a boat. Individuals and groups who hope to foster local resilience in the face of such a future probably ought to make such simple and readily available technologies as solar water heating, solar space heating, home-scale wind power and the like central themes in their planning. Up to now this has rarely happened, and the hold of the logic of abundance on our collective imagination is a good part of the reason why.

The electrical power grid is only one example of a system that only makes sense in a world where energy is so abundant that even huge inefficiencies don't matter. It's hardly difficult to think of others. To think in these terms, though, and to begin to explore more economically viable options for meeting individual and community needs in an age of scarce energy, is to venture into a nearly unexplored region where most of the rules that govern contemporary economic life are turned on their heads.

The End of the Information Age

Probably the best example is the looming impact of a future of energy constraints on the ways that modern industrial nations store, process and distribute information. Few subjects have been loaded with as much hype as this one. Our era, the media never tires of repeating, is the Information Age; economic sectors dealing with mere material goods and services have been relegated to Third World sweatshops, while the economic cutting edge deals entirely in the manufacture, sales and service of information in various forms. As usual — can you think of a short-term trend that hasn't been identified as a wave of the future destined to rise up an exponential curve to infinity, or at least absurdity? I can't — the assumption is that the future will be just like the present, but more so, with more elaborate technologies providing more baroque information products and services as far as the eye (or the webcam) can see.

This is hardly a new vision of the future. In his 1909 novella "The Machine Stops," E.M. Forster provided a remarkably exact dissection of contemporary cyberculture's idea of its destiny most of a century in advance. It's a great story on its own terms, but it also puts a finger on the central weakness of an information-centered society: information does not exist without a physical substrate, and if the physical substrate goes, so does the information.

In Forster's story, that substrate was the Machine — an interconnected global technostructure that provided the necessities and luxuries of life to billions of people who spent their lives in hive-like cells, staring into screens and tapping on keyboards, for all the world like today's computer geeks. Adept at manipulating abstract ideas, the inhabitants of the Machine lost touch with the fact that their universe of information only existed because the physical structure of the Machine kept it there, and their attitude toward the Machine evolved into a religious reverence free of any reference to the practical realities of the Machine's workings. The skills needed to apply tools to pipes and wires dropped out of use, and the consequences — minor malfunctions snowballing into major ones and finally into total systems failure and mass death — followed from there.

Now of course fiction is fiction, and the events that caused Forster's Machine to stop are unlikely to be repeated in the real world. The central concept, though, demands attention, because our Machine — the Internet — depends just as much on a physical substrate as the one in Forster's novella. In our case, that substrate is the global network of communications links and server farms, and the even vaster economic and technical infrastructure that keeps them funded, powered and supplied with the trained personnel and spare parts they need.

Very few people realize just how extravagant a supply of resources goes to maintain the information economy. The energy cost to run a home computer is modest enough that it's rarely noticed, for example, that each one of the big server farms that keep today's social websites up and running use as much electricity as a midsized city. Multiply that by the tens of thousands of server farms that keep today's online economy going, and the hundreds of other energy-intensive activities that go into maintaining the Internet and manufacturing the equipment it uses, and it may start to become clear how much energy goes into putting pretty pictures and text onto your computer screen.

It's not accidental that the Internet came into existence during the last hurrah of the age of cheap energy, the quarter century be-

tween 1980 and 2005 when the price of energy hit the lowest levels
in human history. Only when energy was quite literally too cheap
to bother conserving could so energy intensive an information network be constructed. The problem with this, as already discussed,
is that the conditions that made energy cheap and abundant during that quarter century have already come and gone, and the economics of the Internet take on a very different shape as energy
becomes scarce and expensive again.

Let's start with the law of supply and demand. As we have seen,
this has its problems when applied to energy and other primary
goods, but it works very well when applied to nonessential services such as the Internet. If the cost of maintaining the Internet
in its current form outstrips the income that can be generated by
it, the Internet becomes a losing proposition, and cheaper modes
of information storage and delivery will emerge to replace it. Governments may decide to maintain some form of Internet as long
as they can, even when it becomes an economic sink, but this does
not mean that everyone in the industrial world will have the same
access they do today.

Instead, as energy costs move unsteadily upward and resource
needs increasingly get met — or not — on the basis of urgency, we
can expect access costs to rise, government regulation to increase,
Internet commerce to be subject to increasing taxation and rural
regions and poor neighborhoods to lose Internet service altogether. There may well still be an Internet a quarter century from
now, but it will likely cost much more, reach far fewer people and
have only a limited resemblance to the free-for-all that exists today.
Newspapers, radio and television all moved from a growth phase
of wild diversity and limited regulation to a mature phase of vast
monopolies with tightly controlled content; even in the absence
of energy limits, the Internet would be likely to follow the same
trajectory, and the rising costs imposed by the end of cheap energy
bid fair to shift that process into overdrive.

As the economic burden of the Internet's immense energy usage begins to bear down, then, other technologies less dependent on huge energy inputs will become more economical, driving a spiral in which rising costs and restricted access will cut into Internet service while simpler technologies absorb a growing range of its current economic roles. Finally, once economic contraction has gone far enough, the Internet may just simply drop out of use altogether because the economic basis for its operation will have gone away.

It's true, of course, that the Internet could be operated more efficiently than it is today. Efforts to increase efficiency, however, are subject to a law of diminishing returns; a range of limits ultimately rooted in thermodynamics put a ceiling on just how efficient any process can get. Such gains also have costs of their own: research and development do not come cheap, nor do the construction and installation of more efficient equipment, and the budget cuts currently sweeping through companies and universities worldwide—themselves the harbingers of much greater cuts to come—hardly support the act of faith that claims infinite technological improvement will be the answer to this and all other problems.

Nor is it valid to put the possibility of increased efficiency for the Internet on one side of the balance and ignore the equivalent possibilities on the other side. After all, other technologies—some of which already use less energy than the Internet—are just as liable to see gains in efficiency as the Internet. Even a more efficient Internet is unlikely to be the most economical way to use the sharply constrained energy and resource flows of the future; if another technology or suite of technologies can provide something like the same services at a lower cost, that technology or suite of technologies will outcompete the Internet. Thus if it costs less, all things considered, to send messages over shortwave radio, order products by mail from a catalog and get pornography from a local adult bookstore, then the Internet will fall by the wayside, or at best will be propped up for non-economic reasons as long as economic realities make it possible to do so.

The entire supply chain that keeps the Internet and its potential competitors running has to be factored into these calculations. It's easy to see the Internet as uniquely efficient if all you take into account is the energy going into your home computer. The gigawatts used by server farms are not the only unnoticed energy that goes into the Internet, though; putting those gigawatts to work requires an electrical grid spanning most of a continent, backed up by the immense inputs of coal and natural gas that put electricity into the wires, and a network of supply chains that stretches from coal mines to power plants to the oil wells that provide diesel fuel for trains and excavation machines. The server farms draw on a vast array of supporting services and manufacturers, from the overseas mines that produce rare earths for semiconductor doping through the factories that turn out components to the colleges that turn out trained technicians — and the list goes on.

All told, a fair fraction of the world's industrial economy helps support the Information Age in one way or another, and many of those support functions can't be done at all in a less centralized way or at a lower level of technology. Most of the potential replacements for the Internet don't suffer from that limitation. It's entirely possible to build a shortwave radio by hand, for example, using components that can be built from readily available materials — there are radio amateurs alive today who did precisely that before the postwar electronics boom made manufactured components cheap and easily accessible — and a shortwave radio can communicate across continental differences just as effectively as the Internet does. As the cost of concentrated energy becomes a major economic burden, these differences will matter and give a massive economic advantage to less energy-intensive ways of accomplishing things.

The Economics of Contraction

To suggest that the Internet will turn out to be, not the wave of the future, but a relatively short-lived phenomenon on the crest of the age of cheap abundant energy, is to risk running headlong into the logic of abundance already discussed. It's essential not to get caught up in thinking of how many advantages the Internet might provide to a post-abundance world, because the advantages conferred by the Internet in no way mean that it must continue to exist. The fact that something provides an advantage does not guarantee that it is economically viable.

An example from one of the most famous cases of social collapse is relevant here. On Easter Island, the native culture built a thriving society that got most of its food from deepwater fishing, using dugout canoes made from the once-plentiful trees of the island. As the population expanded, however, the demand for food expanded as well, requiring more canoes, along with many other things made of wood. Eventually, the result was a deforestation so extreme that all the tree species once found on the island went extinct. Without wood for canoes, deepwater food sources were out of reach, and Easter Island's society imploded in a terrible spiral of war, starvation and cannibalism.[2]

It's easy to see that nothing would have offered as great an economic advantage to the people of Easter Island as a permanent source of trees for deepwater fishing canoes. It's just as easy to see that once deforestation had gone far enough, nothing on earth could have provided them with that advantage. Well before the final crisis arrived, the people of Easter Island—even if they had grasped the nature of the trap that had closed around them— would have faced a terrible choice: leave the last few big trees standing and starve today, or cut them down to make canoes and starve later on. All the less horrific options had already been foreclosed.

Further back in Easter Island's history, when it might still have been possible to work out a scheme to manage timber production sustainably and produce a steady supply of trees for canoes, doing so would have required harsh tradeoffs: one additional canoe per year, for example, might have required building or repairing one less house each year. Both the canoe and the house would have yielded significant economic advantage, but it wouldn't have been possible to get both. In a world of limited resources, in other words, it's not enough to insist that a given allocation of resources has economic advantages; you must also show that the same resources would not be better used in some other way or for some other need.

This sort of thinking will inevitably be applied to technology in a future of energy and resource constraints, and any attempt to make sense of that future in advance needs to start by embracing that sort of thinking now. One useful way to assess the vulnerability of any current technology in the post-abundance world, in fact, is to note the difference between the direct and indirect energy inputs needed to keep it working and the inputs needed for other, potentially competing technologies that can provide some form of the same goods or services. All other factors being equal, a technology that depends on large inputs of energy will be more vulnerable and less economically viable in an age of energy scarcity than a technology that depends on smaller inputs, and the larger the disparity in energy use, the greater the economic difference. In turn, communities, businesses and nations that choose less vulnerable and more economical options will prosper at the expense of those that do not, leading to a generalization of the more economical technology. It really is that simple.

You might think that this sort of economic analysis would be an obvious and uncontroversial part of peak oil planning. Of course it's nothing of the kind. Most discussion and planning around the subject of energy futures these days pays no more than lip service to economics, if it deals with that dimension at all, and a great many of the plans being circulated these days look very appealing until you do the math and discover that the most basic questions about resource inputs and economic outputs haven't been addressed.[3]

Part of this blindness to the economic dimension is hardwired into contemporary culture. It hardly needed the mass exodus into delusion that drove the recent real estate bubble to prove that most people in the industrial world nowadays think that getting something for nothing is a reasonable business plan. We have lived with so much abundance for so long that a great many of us seem to have lost any sense that there are limits we can't borrow or bluster our way around.

To a very great extent, indeed, the last 300 years of economic expansion have been driven by a borrowing binge even more colossal, and ultimately more catastrophic, than the one that began its implosion in 2008. The difference is that instead of borrowing from banks we borrowed from the Earth's stockpile of fossil carbon, and squandered most of our borrowings on vaster equivalents of the salad shooters and granite countertops that absorbed so much fictitious value during the late boom. By the time Nature's collection agencies get through with us, they may have repossessed everything we bought with our borrowings — which is to say nearly everything we've built over the last three centuries.

This wholesale contraction will pose an additional challenge to the future, because nearly all our recent technologies displaced older technologies that provided the same services on a more sustainable basis. Here again, the Internet provides a solid example. The collapse of the newspaper industry is one widely discussed example of the Internet's toll on older and more sustainable information media, but another — the death spiral of American public libraries — is likely to have a much wider impact in the decades and centuries to come.

Among the most troubling consequences of the current economic crisis are wholesale cuts in state and local government funding for libraries. Some of the proponents of these budget cuts have been caught in public insisting that with the rise of the Internet nobody actually needs public libraries any more. Of course public libraries provide many services the Internet doesn't, and provides them to all those people who can't afford Internet access. The crucial point here, though, is that the public library will still be a viable information technology in a post-petroleum society. When Ben Franklin founded North America's first public library, after all, he did it without benefit of fossil fuels.

If public libraries can be kept open during the waves of economic crisis that will punctuate the end of the age of abundance, everyone will likely be the better for it. I am sorry to say that this is probably not the way things will fall out. The current wave of library downsizing is very probably a harbinger of things to come: pressed between too many demands and too little funding to go around, library systems — like public health departments and a great many other institutions that make community life viable — are far too likely to draw the short straw. Exactly this sort of short-term thinking drove the loss of vast amounts of information and cultural heritage in the collapse of past civilizations.

As we move into the penumbra of the deindustrial age, then, it's crucial to start thinking about the options open to us — individually and collectively — with an eye toward their long-term viability and to the hard reality of a world of ecological limits. When today's data centers are crumbling ruins long since stripped of valuable salvage, and all the data once stored there has gone wherever magnetic patterns go to when they die, the thinking that led politicians to gut library systems on the assumption that the Internet will take up the slack will look remarkably stupid. Still, the habits of thought instilled by the age of cheap abundant energy are hard to shake off, and from within them such mistakes are hard to avoid.

The Twilight of the Machine

The end of the age of cheap abundant energy thus brings with it an unavoidable reshaping of our most basic ideas about economic development. For the last three centuries or so, as already discussed, the effective meaning of this phrase has centered on the replacement of human labor by machines powered by fossil fuel. All the other measures of development — and, of course, plenty of them have been offered down through the years — either reflect or presuppose that basic economic shift.

The replacement of labor with mechanical energy has even come to play a potent role in the popular imagination. From the machine-assisted living of *The Jetsons* to the darker image of reality itself as a machine-created illusion in *The Matrix*, the future has come to be defined as a place where people do even less work with their own muscles and minds than they do today. All this is the product of the logic of abundance: the notion, rooted down at the core of industrial society's worldview, that there will always be enough resources to let people have whatever it is that they think they want.

Abandon that comfortable but unjustifiable assumption, and the future takes on a very different shape. In a world where everything but human beings will be in short supply, it makes no sense whatever to deploy increasingly scarce resources to build, maintain and power machines to do jobs that human labor can do equally well. An example may be useful here. Let's take Rosie the Riveter, the iconic woman factory worker of Second World War fame, and match her up against one of the computer-guided assembly line robots that have replaced so many workers in production lines in the industrial world. We might as well pit icon against icon and call the robot HAL 9000.

Both of them serve the same economic function, we'll assume, riveting parts together on an assembly line. It's a credo of contemporary economics that HAL is more productive than Rosie, but since the term "productivity" in contemporary economic parlance means "labor productivity," or, in other words, how much production you can get per worker, any machine is by definition more productive than human labor. In a world of resource constraints, though, this definition of productivity becomes very hard to justify. It may be true that HAL can work long shifts at all hours with only the very occasional break for maintenance — at least this is what the robot salesman will tell you — and Rosie cannot. Still, in a world of resource scarcity, Rosie has a crucial advantage that more than offsets HAL's capacity for night shifts: her operating requirements are much less energy- and technology-intensive to meet than HAL's.

We can start with the energy source used by each riveter. HAL requires electricity — quite a bit of it, within fairly tight specifications of voltage, amperage and cycles per second. For her part, Rosie requires food, and though she's been known to take a second helping in the factory cafeteria, her fuel needs are fairly modest compared to those of the machine. Her tolerances for variability in energy sources are also much broader than HAL's — if you have trouble believing this, spend a few minutes paging through an old wartime cookbook.

HAL's maintenance requirements are just as exacting. He needs lubricants that meet precise specifications and an assortment of spare parts ranging from zinc bushings to integrated circuits, none of which he can provide for himself. All of them must be manufactured off-site, and some (such as the integrated circuit) cannot be made without extremely expensive, complex facilities demanding intricate technological infrastructures of their own. Rosie's maintenance needs, by contrast, involve little more than eight hours of sleep and a modest additional amount of food. ("I'll have two scoops of slumgullion today, Franny, thanks. It's been a hard shift.")

When it's necessary to replace HAL, a huge array of industrial facilities — mines, smelters, chemical plants, chip fabrication plants and one or (usually) several factories—have to be brought into play to produce HAL 9100. Unlike HAL, Rosie can manufacture her own replacement, and while it will take most of two decades before Rosie Jr. is ready to tie her hair up in a bandanna and take her place on the assembly line, Rosie's own working life is longer still, so the replacement cycle is not a problem for her. In a world with nearly seven billion people on it, of course, it's hardly necessary to wait for Rosie herself to reproduce in order to find a new riveter, or ten thousand of them.

Finally, what happens if the economy changes so that there's no longer a need for as many riveters, as happened at the end of the Second World War? It might be possible to retool HAL for some other industrial process, but for reasons of efficiency, most assembly line robots are designed for a very limited range of operations and get mothballed or go to the scrap heap when the demand for their services goes away. Rosie, on the other hand, is capable of a nearly limitless range of productive economic activities and can head off to some other career when the factory closes down, leaving HAL to sing "Daisy Bell" to himself on the deserted assembly line.

These arguments could be developed at even greater detail, and with less whimsy, but I trust the point has been made: HAL's appearance of greater productivity depends on access to a support system of factories and services vastly larger than the one Rosie needs, and his support system in turn depends on the availability of cheap abundant energy and a wide range of specialized resources and supplies that hers does not require. What makes HAL more economical in an age of resource and energy abundance is the low cost of that fossil fuel energy. During an age of resource scarcity the equation changes completely, because the goods and services that support Rosie can be produced with a much simpler technology, and with much less in the way of concentrated energy, than the goods and services that support HAL.

There's a reason for this, of course: human beings, along with all other living beings, evolved over millions of years in a world of energy and resource scarcity. Our hominid ancestors, and all their ancestors down the lineage of evolution all the way to those first prokaryotic cells back in the dank Archean mists, spent most of their lives confronting the hard logic of Malthus by which population rises right up to the limits of carrying capacity. There are some multicellular organisms that have requirements as exacting and purposes as limited as most machines, but there are not many, and our species ranks right up there with rats, crows and cockroaches among Nature's supreme generalists.

It's only in the highly atypical conditions of the last three centuries, then, that machines are more economical than human laborers. This is why, for example, nobody in the Roman world used Hero of Alexandria's aeolipile, the first known steam engine, as a source of power for industry or transport. Craft traditions in the Roman Empire would certainly have been up to the challenge, and the aeolipile was much discussed at the time as an interesting curiosity; what was lacking was the recognition that the black gooey stuff that seeped from the ground in certain places, or the black flammable stone we call coal, could be used in large quantities as fuel for such a machine. Lacking that, in turn, the aeolipile was never more than a curiosity, for the fuel supplies the Roman world knew about were already committed to existing economic sectors, while human and animal muscle were abundant, familiar and cheap.

As the industrial age winds down, human muscle will again be abundant. Will it be cheap? Almost certainly, yes — and that means that real wages for most people in the industrial world will continue their current slide toward Third World levels. I wish I could say otherwise, not least because my chances of taking part in that slide are tolerably high. Still, part of what has made the last three centuries so atypical is the extent to which ordinary people in the industrial world have been able to rise out of the hand-to-mouth existence typical of most of humanity for most of history and partake of a degree of comfort and security. That state of affairs could never have been permanent, because it was made possible only by using up fantastic amounts of fossil sunlight at a pace so extravagant that the quest to figure out what to do with all that energy has been a major driver of economic change for more than a century now. It's simply our bad luck to live at a time when the bill for all that extravagance is coming due.

All this should be fairly straightforward and uncontroversial. It isn't, of course, because the contemporary faith in the superiority of the machine reaches deep into the irrational levels of our collective psyche. When Lewis Mumford titled one of his most significant works *The Myth of the Machine* he was not engaging in hyperbole. The thought that Rosie the Riveter could go head to head with HAL 9000 and win is unthinkable to most of us; it's a matter of folk belief throughout industrial society that the machine always wins, or at least that any victory over it is as temporary and fatal as John Henry's Pyrrhic triumph over the steam drill.

The machine is our totem, the focus of a great deal of our culture's sense of value and purpose, and most people in the industrial world accord it the same omnipotence that older religions claim for their gods. The sheer volume of popular culture over the last century or so that fixates on the notion of machines taking over the world and treating humanity the way industrial humanity has so often treated other living things, is one indicator of the mythic power machines have come to hold in our collective imagination. It's for this reason, I think, that so many of us simply can't imagine a future in which machines will be less economically viable than human labor.

Yet if it costs the equivalent of five dollars a day to hire a file clerk and a secretary at post-abundance wage scales, and it costs the equivalent of ten dollars a day in expensive and unreliable electricity to run a computer to do the same things, those businesses that hope to succeed will hire the file clerk and the secretary, and the computer will be left to gather dust. Now it's true, as fans of computers are quick to point out, that computers will do things that secretaries and file clerks can't, but the reverse is also true — try asking your computer sometime to go pick up takeout lunch for the office from a place that doesn't deliver — and many of the abilities unique to computers are conveniences rather than necessities; businesses got along very well without them for thousands of years, remember.

The technology that's useful to help a human worker do his job more effectively is not the same as the technology that's needed to replace him or her with a machine. As cheap abundant energy becomes a thing of the past, replacing workers with machines will no longer be a viable option, but providing workers with tools that will make their labor more productive is quite another matter. The problem here is that very few people are used to thinking in these terms. While every industry in the world once had a vast amount of practical knowledge about the tools and training human workers needed to do their jobs well, nearly all of that knowledge is endangered, if it hasn't already been lost.

Consider the slide rule, as one example among many. Until the 1970s, it was the engineer's inseparable companion; every technological advance from the mid-nineteenth century until Apollo 11 landed on the Moon was made possible, in part, by competent manipulation of this simple, flexible, ingenious analog computer by people who knew how to make the most of its strengths and work within its limits. Since it doesn't require a massive and technologically complex support structure to construct, maintain and operate them — any good cabinetmaker can make one, and their proper fuel is a scoop of the same slumgullion that kept Rosie going on her shift — slide rules are likely to be just as useful on the downslope of the industrial age as they were on the way up. If, that is, anybody on Earth still remembers how to use one when we get to that point along the curve of economic contraction.

This is where the myth of the machine — the conviction, as irrational these days as it is pervasive, that the best person for any job is always not a person at all, but a machine — stops becoming a curious twist of our collective imagination and turns into a trap we ignore at our peril. As peak oil moves closer to center stage in the historical drama of our time, and makes the gargantuan technostructure we've built on a foundation of cheap abundant energy ever more problematic to sustain, the most common response from the centers of power and the masses alike is to call for the development of even more complex, gargantuan and tightly interlinked machines, pushing the technostructure in the direction of greater risk and greater dysfunction. It's hardly an exaggeration to suggest that if it turned out we were all about to perish en masse from building too many machines, the first reaction of most people in today's industrial cultures would likely be to insist that the answer was to build more machines.

Thus we will doubtless see plenty of shiny new machines built in the years to come, and they will doubtless do their fair share and more to push industrial civilization further down the arc of its decline. As the ancient Greeks knew well, it's the essence of tragedy that the *arete*, the particular excellence, of a tragic hero also turns out to be his *hamartia* or fatal flaw. Put another way, a civilization that lives by the machine can expect to die by the machine as well.

Toward Appropriate Technologies

The heretical minority that has learned to mistrust the myth of the machine, however, may well keep in mind that as the age of scarcity dawns, educating people is a far more useful project than building machines, and doing as much as possible to insure that individuals, families and communities have the skills and simple tools they need to work productively may just be the most promising response to the future ahead of us. There's nothing really remarkable about that future; it's simply that the unparalleled abundance that our civilization bought by burning through half a billion years of stored sunlight in three short centuries has left most people in the industrial world clueless about the basic realities of human life in more ordinary times.

It's this cluelessness that underlies so many enthusiastic discussions of a green future full of high technology and relative material abundance. Those discussions also rely on one of the dogmas of the modern religion of progress: that the accumulation of technical knowledge was what gave the industrial world its three centuries of unparalleled wealth. Since technical knowledge is still accumulating, the belief goes, we may expect more of the same in the future. Now in fact the primary factor that drove the rise of industrial civilization, and made possible the lavish lifestyles of the recent past, was the recklessness with which the earth's fossil fuel reserves have been extracted and burnt over the last few centuries. The explosion of technical knowledge was a consequence of that, not a cause.

In what we might as well get used to calling the real world— that is, the world as it is when human societies don't have such immense quantities of highly concentrated energy ready to hand—the primary constraints on the production of wealth are the natural limits imposed by the annual yield of energy resources and raw materials from the primary economy. Even after two billion years of evolution, photosynthesis only converts less than one percent of the solar energy falling on leaves into chemical energy that can be used for other purposes, and that only when other requirements—water, soil nutrients and so on—are also on hand. Other than a little extra from wind, running water, sunlight and other naturally occurring energy sources, that trickle of energy from photosynthesis is what a nonindustrial society has to work with; that's what fuels the sum total of human and animal muscle that works the fields, digs the mines, wields the tools of every craft and does everything else that produces wealth. This, in turn, is why most people before the industrial age had so little: the available energy, and the other resources that can be extracted and used with that energy, are too limited to provide any more.

The same limits apply to today's nonindustrial nations, though for different reasons. Here the problem is the assortment of colonial and neocolonial arrangements that drain most of the world's wealth into the coffers of a handful of industrial nations, and leave the rest to tussle over the little that's left. The five percent of the world's population that happens to live in the United States, for example, doesn't get to use roughly a third of the world's resources and industrial production because the rest of the world has no desire to use a fairer share themselves. Rather, our prosperity is maintained at their expense, and until recently — when the current imperial system began coming apart at the seams — any nonindustrial country that objected too loudly to that state of affairs could count on having its attitude adjusted by way of a coup d'état or "color revolution" stage-managed by one or more of the powers of the industrial world, if not an old-fashioned invasion of the sort derided in Tom Lehrer's ballad "Send the Marines."

Over the last century or so, a handful of insightful thinkers have tried to explore ways in which the cycle of exploitation and dependency can be broken. Schumacher is perhaps the most important of these. Even though he brought his substantial background as a working economist to bear on the problems facing the world's nonindustrial nations. He drew his central ideas from a different source: the largely neglected economic ideas of Gandhi.

Most people in the industrial world think of Mohandas K. Gandhi as a spiritual leader, which of course he was, and as a political figure, which he also was. It's not as often remembered that he spent quite a bit of time and effort developing an economic theory appropriate to the challenges facing a newly independent

India. His suggestion, to condense some very subtle thinking into too few words, was that a nation that had a vast labor force but very little money was wasting its time investing that money in state-of-the-art industrial plants. Instead, he suggested, the most effective approach was to equip that vast labor force with tools that would improve their productivity within the existing structures of resource supply, production and distribution. Instead of replacing India's huge home-based spinning and weaving industries with factories, for example, and throwing millions of spinners and weavers out of work, he argued that the most effective use of India's limited resources was to help those spinners and weavers upgrade their skills, spinning wheels and looms, so they could produce more cloth at a lower price, continue to support themselves by their labor, and in the process make India self-sufficient in textile production.

This sort of thinking flies in the face of nearly every mainstream economic theory since Adam Smith. Since nearly every mainstream economic theory since Adam Smith has played a role in landing the industrial world in its current mess, I'm not so sure this is a bad thing. Current economics dismisses Gandhi's ideas on the grounds of their "inefficiency," but this has to be taken in context; "efficiency" in today's economic jargon means nothing more than efficiency in producing somebody a profit. As a way of keeping millions of people gainfully employed, stabilizing the economy of a poor nation and preventing its wealth from being siphoned overseas by predatory industrial nations, Gandhi's proposal is arguably very efficient indeed — and this, in turn, was what brought it to the attention of E. F. Schumacher.

One of Schumacher's particular talents was a gift for intellectual synthesis. His work is full of cogent insights that sum up a great deal of more specialized work and make it applicable to a wider range of circumstances. This is what he did with Gandhi's ideas. Schumacher argued that talk about "developing the Third

World" typically neglected to deal with one of the most pragmatic issues of all—the cost of setting up workers with the tools they needed to work.[4]

An example will help explore his logic. You are the president of the newly independent nation of Imaginaria. You've got a population that's not particularly well fed, clothed and housed, and a high unemployment rate; you've got a very modest budget for economic development; your nation, however, has raw materials of various kinds, which could be used to feed, clothe and house the Imaginarian people. Your foreign economic advisers, who not co-incidentally come from the industrial nation that used to be your country's imperial overlord, insist that your best option is to use your budget to build a big modern factory that will turn those raw materials into goods for export to the industrial nation by its own merchants, giving your country cash income to buy goods from them, and in the process employing a few thousand Imaginarians as factory workers.

Not so fast, says Schumacher. If your goal is to feed, clothe, house and employ the Imaginarian people, building a factory is a very inefficient way to go about it, because that approach requires a very large investment per worker employed. You can provide many more Imaginarians with productive jobs for the same amount of money, by turning to a technology that's less expensive to build, maintain and supply with energy and raw materials—say, by providing them with hand tools and workbenches instead of state-of-the-art fabrication equipment, and setting up supply chains that supply them with local raw materials instead of imports from abroad. The goods those workers produce may not be suitable for export, but that's not necessarily a problem—remember, your goal is to feed, clothe and house Imaginarians so they will reelect you, and maximizing production for domestic use will make you more popular in the Imaginarian street, since less of the value produced by those workers will be skimmed off by the middlemen who manage international trade. Furthermore, since your nation won't have to trade with overseas producers for as many of the necessities of life, the need for cash from overseas goes down, and you get an economy less vulnerable to foreign exchange shocks into the bargain.

This was the basis for what Schumacher called "intermediate technology," and the younger generation of activist-inventors who followed in his footsteps called "appropriate technology." Their idea was that relatively simple technologies, powered by locally available energy sources and drawing on locally available raw materials, could provide paying jobs and an improved standard of living for working people throughout the Third World. A lot of productive thinking went into these projects, and there were impressive success stories before the counterrevolution of the 1980s cut what little funding the movement had been able to find. Mind you, Schumacher's thinking was never popular among economists or the business world, and it happened more than once that countries that tried to adopt such economic policies were treated to the sort of attitude adjustments mentioned earlier. Still, pay attention to those Third World nations that have succeeded in becoming relatively prosperous and you'll find that some version of Schumacher's scheme played a significant role in helping them do that.

It's when the same logic is applied to the industrial world, though, that Schumacher's ideas become most relevant to the project of this book. As the economies of the industrial world begin to take on features until recently confined to the nonindustrial world, thinking designed for the nonindustrial world may be a good deal more applicable here and now than the conventional wisdom might suggest. It seems utterly improbable that the governments of today's industrial powers will have the foresight or, for that matter, the common sense, to realize that economic policies that deliberately increase the number of people earning a living might be a very good idea in an age of pervasive structural unemployment—

or, for that matter, to glimpse the unraveling of the industrial age, and realize that within a finite amount of time, the choice will no longer be between high-tech and low-tech ways of manufacturing goods, but between low-tech ways and no way at all. Still, national governments are not the only players in the game.

What Schumacher proposed, in fact, is one of the missing pieces to the puzzle of sustainability in an age of industrial decline. The economies of scale that made centralized mass production possible in recent decades were simply one more side effect of the vast amount of energy the industrial nations burnt up during that time. As fossil fuel depletion brings those excesses to an end, the energy and other resources needed to maintain centralized mass production will no longer be available, and what I've described above as the economics of the real world come into play. At that point, the question of how much it costs to equip a worker to do any given job becomes a central economic issue, because any resources that have to go to equipping that worker must be taken away from another productive use.

In the twilight of the age of cheap energy, the most abundant energy source remaining throughout the world will be human labor, and as other resources become more costly, the price of labor—and thus the wages that can be earned by it—will drop accordingly. At the same time, human labor has certain crucial advantages in a world of energy scarcity, just as it did in earlier eras of economic contraction and social decline. Unlike other technologies, human labor is fueled by food, which is a form of solar energy. Our agricultural system produces food using fossil fuels, but this is a habit of an age of abundant energy; field labor by human beings with simple tools, paid at close to Third World wages, already plays a crucial role in the production of many crops in the US, and this will only increase as wages drop and fuel prices rise.

The agriculture of the future, like agriculture in any thickly populated society with few energy resources, will thus use land intensively rather than extensively, rely on human labor with hand tools rather than more energy-intensive methods and produce bulk vegetable crops and relatively modest amounts of animal protein; the agricultural systems of medieval China and Japan, chronicled by F. H. King in *Farmers of Forty Centuries*, are as good a model as any. Such an agricultural system will not support seven billion people, but then neither will anything else, and a decline in population as malnutrition becomes common and public health collapses is a fairly safe bet for the not-too-distant future.[5]

For similar reasons, the economies of the future will make use of human labor, rather than any of the currently fashionable mechanical or electronic technologies, as their principal means for getting things done. Partly this will happen because in an overcrowded world where all other resources are scarce and costly, human labor will be the cheapest resource available, but it draws on another factor as well.

This was pointed out many years ago by Lewis Mumford in *The Myth of the Machine*. He argued that the revolutionary change that gave rise to the first urban civilizations was not agriculture, or literacy or any of the other things most often cited in this context. Instead, he proposed, that change was the invention of the world's first machine — a machine distinguished from all others in that all of its parts were human beings. Call it an army, a labor gang, a bureaucracy or the first stirrings of a factory system; in these cases and more, it consisted of a group of people able to work together in unison. All later machines, he suggested, were attempts to make inanimate things display the singleness of purpose of a line of harvesters reaping barley or a work gang hauling a stone into place on a pyramid.

This kind of machine has huge advantages in an world of abundant population and scarce resources. It is, among other things, a very efficient means of producing the food that fuels it and the other items needed by its component parts, and it is also very efficient at maintaining and reproducing itself. As a means of turning solar energy into productive labor, it is less efficient than current technologies, but its simplicity, its resilience and its ability to cope with widely varying inputs give it a potent edge over these in a time of turbulence and social decay.

This kind of machine, it deserves to be said, is profoundly repellent to many people in the industrial world, doubtless including many of those who are reading this book. It's interesting to think about why this should be so, especially when some examples of the machine at work—Amish barn raisings come to mind—have gained iconic status in the alternative scene. It is not going too far, I think, to point out that the word "community," which receives so much lip service these days, is in many ways another word for Mumford's primal machine. For the last few centuries, we have tried replacing that machine with a dizzying assortment of others; instead of subordinating individual desires to collective needs, like every previous society, we have built a surrogate community of machines powered by coal and oil and natural gas to take care, however sporadically, of our collective needs. As those resources deplete, societies used to directing nonhuman energy according to scientific principles will face the challenge of learning once again how to direct human energy according to older and less familiar laws. This can be done in relatively humane ways, or in starkly inhumane ones; what remains to be seen is where along this spectrum the societies of the future will fall.

Becoming a Third World Country

Recognizing that the ideas Schumacher crafted for the newly independent nations of his own time are increasingly relevant to the industrial nations of the present, however, is only one part of a broader and even less welcome realization: the recognition that perhaps the best way to describe the changes ahead is to say that most of the world's industrial nations are in the process of becoming Third World countries.

For all its current power and wealth, the United States is firmly in the lead of that transition. What distinguishes the world's non-industrial countries from the privileged industrial minority of the world's nations? Third World nations import most of their manufactured goods from abroad and export mostly raw materials; that's been true of the United States for decades now. Third World economies have inadequate domestic capital and are dependent on loans from abroad; that's been true of the United States for just about as long. Third World societies are economically burdened by severe problems with public health; the United States ranks dead last for life expectancy among industrial nations, and its rates of infant mortality are on a par with those in Indonesia, so that's covered. Third World nation are very often governed by kleptocracies that run the affairs of government for the private benefit of the few, and a strong case can be made that this, too, is currently the state of affairs in the United States.

There are, in fact, precisely two things left that differentiate the United States from any other large, overpopulated, impoverished Third World nation. The first is that the average standard of living in the United States, measured either in money or in terms of energy and resource consumption, stands well above Third World levels — in fact, it's well above the levels of most other industrial nations. The second is that the United States has the world's most expensive and technologically complex military. Those two factors are closely related, and understanding their relationship is crucial in making sense of the end of the "American Century" and the decline of the United States, along with other industrial nations, to Third World status.

The US has the world's most expensive military because, just now, it has the world's largest empire. It's not considered polite to talk about America's global hegemony in those terms, but the US does not keep its troops garrisoned in 140 countries around the world for the sake of their health. The US empire functions, like other empires, as a way of tilting economic relationships between nations so that wealth flows out of the rest of the world and into the coffers of the imperial nation. It never occurs to most Americans to wonder why it is that the five percent of the world's population who live in the US get to use around one third of the world's production of natural resources and industrial products; the economics of empire are the reason why.

A century ago, in 1910, Britain had the global empire, the worldwide garrisons, and the torrents of wealth flowing from around the world to boost the British standard of living at the expense of everyone else. A century from now, in 2110, if the technology to maintain a worldwide empire still exists — and it can be done with a much simpler technological basis; Spain managed to build a world empire with no technology more complex than wooden sailing ships and crude cannons — somebody else will be in that position. It won't be America, because empire is the methamphetamine of nations; in the short term, the effects feel great, but in the long term they're lethal. Britain managed to walk away from its empire without catastrophe because the United States was ready, willing and able to take over, and give Britain a place in the inner circle of US allies into the bargain. Most other nations have paid for their imperial overshoot with a century or two of economic collapse, political chaos and social disintegration.

The economics of empire also explain why it is that so many people in the industrial world believe that it makes sense for them to consume much more in the way of goods and services than they produce. During much of the industrial age, a good fraction of Europe did get away with consuming more than it produced by the simple expedient of owning most of the rest of the world and exploiting it for their own economic benefit. As late as 1914, the vast majority of the world's land surface was either ruled directly from a European capital, occupied by people of European descent or dominated by European powers through some form of radically unequal treaty relationship. The accelerating drawdown of fossil fuels over the last three centuries shifted the process into over-drive, allowing the minority of the Earth's population who lived in Europe or the more privileged nations of the European diaspora—the United States first among them—not only to adopt what were, by the standards of all other human societies, extravagantly lavish lifestyles, but to expect that those lifestyles would become even more lavish in the future.

I don't think that more than a handful of people in the industrial world have yet begun to deal with the hard fact that those days are over. European domination of the globe, the first source of Euro-American wealth and power, came apart explosively in the four brutal decades between 1914, when the First World War broke out, and 1954, when the fall of French Indochina put a period to the age of European empire. The United States, which inherited what was left of Europe's imperial role, never achieved the level of global dominance that European nations took for granted until 1914; compare the British Empire, which directly ruled a quarter of the Earth's land surface, with the hole-and-corner arrangements that allow America to maintain garrisons in other people's countries around the world today, and the difference is hard to miss. Now the second and arguably more important source of Euro-American wealth and power—the exploitation of half a billion years of prehistoric sunlight in the form of fossil fuels—has peaked and entered on its own decline, with consequences that bid fair to be at least as drastic as those that followed the shattering of the Pax Europa in 1914.

Thus the end of the bubble of industrialism is made more complex by the end of the bubble of America's political and economic hegemony over the rest of the planet. Like every other empire, we have a tribute economy. We dress it up in free-market clothing by giving our trading partners mountains of worthless paper in return for the torrents of real wealth that flow into the US every day; but the result, now as in the past, is that the imperial nation and its inner circle of allies have a vast surplus of wealth sloshing through their economies. This has had massive social impacts, since handing over a little of that extra wealth to the poor and the working class has proven to be a tolerably effective way to maintain some semblance of social order.

This habit has been around nearly as long as empires themselves. The Romans were particularly adept at it—"bread and circuses" is the famous phrase for their policy of providing free food and entertainment to the Roman urban poor to keep them docile. Starting in the wake of the last Great Depression, when many wealthy people woke up to the fact that their wealth did not protect them against bombs tossed through windows, most industrial nations have done the same thing by ratcheting up working class incomes, providing benefits such as old age pensions and providing plenty of free entertainment in such forms as broadcast television.

What passes by the name of democracy in most industrial nations, as a result, is a system in which factions of the political class buy votes from pressure groups by handing out what the American political slang of an earlier day called by the endearing term "pork." An economy capable of expanding the money supply at will provided ample resources for political pork vendors, and the resulting outpouring of pig product formed a rising tide that, as the saying goes, lifted all boats. The problem faced by the United States in this context is the same one that brought repeated economic crises to the British Empire, the Spanish Empire and many other examples in history: the flood of wealth brought into an imperial economy drives up wages, and when wages in the imperial nation rise far enough above those of its neighbors, it stops being profitable to hire people in that nation for any task that can be done outside it.

The result is a society in which those who get access to pork prosper, and those who don't are left twisting in the wind. Arnold

Toynbee, whose monumental study of the rise and fall of empires remains the most detailed examination of the process,[6] calls these latter the "internal proletariat": those who live within a society but no longer share in its benefits and become increasingly disaffected from its ideals and institutions. In the near term, they are the natural fodder of demagogues; in the longer term, they make common cause with the "external proletariat"—those peoples outside the borders whose labor and resources have become essential to the economy of a dominant state, but who receive no benefits from that economy—and play a key role in bringing the whole system crashing down.

That's the corner into which the United States is backing itself. The flood of lightly disguised tribute from overseas, while it made Americans fantastically wealthy by the standards of the rest of the world, also gutted America's domestic economy and created a culture of entitlement that includes all classes from the bottom of the social pyramid right up to the top. As always happens sooner or later, the benefits of empire are failing to keep pace with its rising costs and, in addition, rising demands for imperial largesse from all parts of society are drawing down an increasingly straitened supply of wealth. Meanwhile other nations with imperial ambitions are circling like sharks. The wisest among them know that time is on their side, and that any additional burden that can be loaded onto a drowning empire will hasten the day when it goes under for the third time and they can close in for the kill.

The Twilight of American Empire

This view of the world situation gets little space in the cultural mainstream or, for that matter, in any of the self-proclaimed alternative scenes. The contrast with a century ago is instructive. Many people in late imperial Britain knew well that the empire on which the sun famously never set—critics suggested that this was because God Himself wouldn't trust an Englishman in the dark—had had its day and was itself setting. The lines of Rudyard Kipling's poem "Recessional" simply put in powerful imagery what many people were thinking at that time:

> Far-called, our navies melt away;
> On dune and headland sinks the fire.
> Lo! All our pomp of yesterday
> Is one with Nineveh and Tyre.

You won't find the same sort of historical sense widespread nowadays, though, and the power of the modern world's secular religion of progress has a great deal to do with it. In 1910, the concept of historical decline was on a great many minds; these days you'll hardly hear it mentioned because the belief in history as perpetual progress has become all the more deeply entrenched as the foundations that made the progress of recent centuries possible have rotted away.

The resulting insistence on seeing all social changes through onward-and-upward-colored spectacles has imposed huge distortions on our perceptions of recent events. One good example is the rise and fall of the so-called "global economy" in recent decades. Its proponents portrayed it as the triumphant wave of a Utopian future that would enable everybody to live like middle-class Americans; its critics portrayed it as the equally triumphant metastasis of a monolithic corporate power out to enslave the world. Very few people saw it as the desperate gambit of a failing imperial society that could no longer afford to run its own economy, and was forced to outsource even its most basic economic functions to the poorer nine-tenths of the world. Nonetheless, this last is what it has turned out to be, and several other nations predictably used the influx of capital and technology to build their own industrial sectors, bide their time, and then enter the market themselves and outcompete the very companies and countries that gave them a foot in the door.

More broadly, it seems to have escaped the attention of a great many observers that the day of the multinational corporation is drawing to an end. The struggle over Russia's energy resources was the decisive battle there, and when Vladimir Putin crushed the Western-funded oligarchs and turned control of his country's energy supply back to its government, that battle was settled with a typically Russian sense of drama. The elegance with which China has turned international trade law against its putative beneficiaries is just as typical: a flurry of corporations owned by the Chinese government have spread their operations throughout the world, using the mechanisms of global trade to lay the foundations of a future Chinese global empire, while the Chinese government has efficiently stonewalled any further trade negotiations that would have put Chinese economic interests at home in jeopardy. More recently, China has begun buying sizable stakes in the multinational corporations that so many well-meaning people in the West once thought would reduce the world to vassalage. The day when ExxonMobil is a wholly-owned subsidiary of China National Offshore Oil Corporation may be closer than it looks.

The same biases that make such global changes invisible have impacts at least as sweeping here in North America. Faith in progress, coupled with the tribute economy's culture of entitlement, have made it nearly impossible for anybody in public life to talk about the hard fact that America can no longer afford most of the social habits it adopted during its age of empire. It's impossible to think of an aspect of daily life that will not change drastically as a result. We will have to give up the notion, for example, that most people in the industrial world ought to go to college and get a "meaningful and fulfilling" job of the sort that can be done sitting at a desk. We will have to abandon the idea that it makes any sense to spend a quarter of a million dollars giving an elderly person with an incurable illness six more months of life. We will have to relearn the old distinction between the deserving poor — those who are willing to work and simply need the opportunity, or who have fallen into destitution through circumstances outside their control—and those who are simply trying to game the system.

The great majority of us will get to find out what it's like to make things instead of buying them, even when that means a sharp reduction in quality; to skip meals, or make do with very little, because the money to pay for anything more simply isn't there; and to treat serious illnesses at home because care from a doctor costs too much. I could go on for paragraphs, but I trust you get the idea. All these changes, it needs to be said, would be inevitable at this point even if the industrial world depended on renewable resources and had a stable, sustainable relationship with the planetary biosphere that supports all our lives. The United States has played its recent hands in the game of empire very badly indeed, and it has responded to each loss by doubling down and raising the stakes even higher. This does not offer much hope for a future of stability.

The global context of the crisis, though, also needs to be kept in mind. The industrial world does not depend on renewable resources, and its relationship with the biosphere is leading it straight down the well-worn path of overshoot and collapse. The endgame of American empire, while it would be taking place anyway, has the additional factor of the limits to growth in play. In an alternate world where energy and resource flows could be counted on to remain stable for the foreseeable future, it's quite possible that one of the rising powers might offer America the same devil's bargain we offered Britain in 1942, and prop up the husk of our empire just long enough to take it over for themselves.

As it is, it cannot have escaped the attention of any other nation that something like a quarter of the world's dwindling resource supplies could be available for other countries, if only the United States were to lose the ability to get energy and other resources from outside its own borders. Plenty of nations would profit mightily from such a readjustment, and nothing so unseemly as a global war would necessarily be required to make it happen; to name only one possibility, it's by no means unthinkable that the United States, having manufactured "color revolutions" to order in countries around the world, might turn out to be vulnerable to the same sort of well-organized mob action here at home.

Exactly how things will play out in the months and years to come is anybody's guess. One of the consequences of America's descent into Third World status, though, is that a great many of us may have scant leisure to contemplate global and national issues amid the struggle to keep food on the table and a roof over our heads. In the long run, this shift in focus may have certain advantages; those nations that undergo the transition soonest, and are thus forced to learn how to get by on the modest energy and resource flows available in the absence of fossil fuels, may find that this gives them a head start in making changes that everyone else will have to make in due time. Still, making the most of those advantages will require a different approach to economics, among other things, than most of us have pursued (or imagined pursuing) so far.

Survival Isn't Cost Effective

Economic factors have played a massive role in putting the industrial world in its current predicament, and play an even more substantial role in blocking any constructive attempt to get out of the corner into which we've painted ourselves. There's all too real a sense in which, if modern industrial civilization perishes it will be because the steps necessary for its survival weren't cost-effective enough. Mind you, this can be interpreted in at least two different ways — as a statement of economic theory or as a statement of practical experience — and both of them are relevant to the crisis of the industrial world. Like any other science, economics is a set of hypothetical models that reflect, with more or less exactness, the

188 THE WEALTH OF NATURE

observed behavior of the world. Too often, though, the models get confused with the reality, and understanding suffers.

This has to be kept in mind when trying to make sense of the economic dimension of industrial civilization's decline and fall, because both sides of the equation—the models and the reality—throw up challenges in the way of constructive action. So do economic policies that are based on the models, and thus function at a second remove from the reality. It's true, of course, that current economic theory has lost touch with reality in critical ways, and a revision of some of the basic ideas of modern economics is essential if we're to make sense of our predicament and do anything constructive in response to it. It's equally true that government policies based on today's misguided economic notions have become massive liabilities to societies struggling to deal with today's crisis, and even this late in the game changes in these policies might still do a great deal of good. Still, it's also true that economic factors in the real world, independent of theory, impose hard limits on what can be done.

This has particular applicability to the current crisis of industrial society. The grim scenario traced out in the seminal 1973 study *The Limits to Growth*—still the most plausible map of the future ahead of us, and thus inevitably the most bitterly vilified[7]—is driven by simple economics. An industrial economy pursuing limitless growth in a finite world, that study pointed out, will have to contend with a steadily depleting resource base and rising pollution generated by the process of industrial production. As resources deplete, the cost of keeping them flowing into the economy will increase in real terms, as more labor and capital have to be invested to extract a given amount of each resource; as pollution levels rise, in turn, the costs of mitigating their impacts on public health, agricultural productivity and other core economic factors go up in the same way and for the same reasons. Those costs have to be paid out of current economic output, leaving less and less for other uses, until economic output itself begins to fall and the industrial world begins its terminal decline.

Now it's easy to insist, if you ignore the economic dimension, that a society facing this sort of crisis can save itself by launching a massive program to build nuclear reactors, solar thermal power plants, algal biodiesel or what have you, and of course this sort of claim has seen endless rehashing over the last couple of decades. The problem is that massive programs of this sort pile additional demands on an already faltering economy and resource base. Any such program has to be paid for, after all, and by this I don't mean that money has to be found for it; in today's hallucinatory economic climate, conjuring money out of thin air is easy enough. No, it has to be paid out of current resource stocks and economic output, which are much less flexible and already have to cover the rising costs of resource depletion and pollution. This is the trap hidden in the limits to growth; once those limits begin to bite, the spare economic capacity that would be needed to build a way out of trouble no longer exists.

Nor is it possible to avoid such economic challenges by embracing such options as the "lifeboat communities" proposed so often in recent years. The basic idea seems plausible enough at first glance: to preserve lives and knowledge through the decline and fall of the industrial age by establishing a network of self-sufficient communities in isolated rural areas, equipped with the tools and technology they will need to maintain a tolerable standard of living in difficult times. The trouble comes, as it usually does, when it's time to tot up the bill.

The average lifeboat community project would cost well over $10 million to establish — many would cost a great deal more — and I have yet to see such a project that provides any means for its inhabitants to cover those costs and pay their bills in the years before industrial civilization goes away. The unstated assumption seems to be that as soon as the intrepid residents of such a community move into their solar-heated cohousing units, start up the wind turbines and the methane generators and get to work harvesting tree crops from the permacultured landscaping all around, industrial civilization will disappear in a puff of smoke and take its taxes, debts and miscellaneous expenses with it. Pleasant though the prospect might seem, I am sorry to say that this isn't going to happen.

The residents of any lifeboat community founded today will not only have to come up somehow with the very substantial sums needed to buy the land, build the cohousing units, wind turbines and so on, and plant all that permaculture landscaping; they will also have to earn a living during the long transitional process that leads from the world we inhabit today to the conditions that will pertain at the bottom of the curve of decline. Some awareness of these difficulties may go a long way to explain why, of the great number of lifeboat communities that have been proposed over the last decade or two, the number that have been built can be counted on the fingers of one foot.

Thus there are limits hardwired into our situation by the inflexible realities that surround us, and we have already strayed far enough over those limits that the payback will inevitably be harsh. The alternative path traced by Schumacher's concept of intermediate technology, and the many practical applications of that concept devised in the years since his time, offer one of the few options for easing the burden of that payback. That path has the further advantage, as already mentioned, that it can be put into practice right now by individuals, families, communities and working groups of various kinds; it need not wait on any collective action by government or society. Still, there are some collective changes that might also be worth pursuing, even this late in the game, and these will form the theme of this book's final chapter.

THE ROAD AHEAD

M ANY PEOPLE nowadays assume that the collapse of the current economic order in the industrial world must lead to
mass death and a descent into savagery. This hardly follows. Most
of the world's nations have undergone political and social collapse
at least once in the course of the last century; the process can certainly be traumatic, but it isn't the end of the world. Whatever
crises drive today's industrial order to its end, and whatever national or international traumas supervene until some degree of stability returns, there will be a place for new policies when the future
governments of today's industrial nations, or the governments of
the new political units that emerge from the wreckage, get to work
picking up the pieces.

This does not mean that all of today's problems can be solved
by a change in policies, or for that matter of politicians. An article
of faith in many circles today holds that everything that's wrong in
the world is the fault of the institutions or personalities in charge,
and can be fixed by replacing them with some other set of institutions or personalities. That notion has been tested more thoroughly by history than any other hypothesis in the social sciences

and it has pretty consistently failed. It seems likely to help matters considerably if we accept that people will not behave like angels no matter how (or whether) they are governed, or who (if anyone) does the governing and if, in the process, we admit that human beings are incurably human — that is, capable of the full spectrum of good and evil all by themselves — rather than moral puppets who can be expected to dance on command to the tune of a good or evil system.

It's easy to come up with a perfect system of human society, in other words, so long as it doesn't have to work in the real world, and it's very easy to compare the glories of a perfect system as it exists on paper to the failings of a system in the real world. Nearly always, though, what John Kenneth Galbraith said about innovation in finance is just as true of innovation in political and social institutions: what gets ballyhooed as new and revolutionary thinking is normally the repetition of a fairly small set of fallacies that worked very poorly the last dozen or so times they were tried, and will work just as poorly this time, too.[1]

Those systems that function at all are fairly few in number, though there are plenty of variations on the basic themes, and the ones that have become standard in the world's industrial nations — representative democracy in politics, a mixed market system in economics — have certain advantages. Though the current examples of these systems are troubled, corrupt and at risk of being overwhelmed by the consequences of bad decisions made over the last few decades, they have still shown themselves to be less dysfunctional most of the time than most of the alternatives that have actually been tried.

There's a reason for this. Though they are rarely recognized as such, societies and economies are natural systems; they are as natural to human beings as hives are to bees and packs to wolves, and the processes of evolution and selection that have shaped them over time are much the same as the processes that shape the social behaviors of other animals. One of the crucial lessons today's environmental crisis is teaching us is that natural systems do not take well to being restructured from the ground up to fit human notions. The evidence of history suggests that with societies, as with other natural systems, change happens most successfully by way of simple mutations rather than complete reshapings. The expansion of civil rights over the last century and a half is a case in point; a set of rights and responsibilities that worked well when applied to a relatively small group of people — in the case of the United States, white men who owned property — has proven to work even better when extended to the adult population as a whole. Meanwhile, attempts to overturn the entire structure of society at once — for example, the French and Russian Revolutions — failed to produce anything but an explosion of violence and tyranny.

Thus no grandiose plan for the complete transformation of everything, of the sort that have been in vogue among social reformers for so many years, will be presented here. Instead, I will suggest a handful of limited, tightly focused changes — mutations, if you will — that could have a real chance, if implemented, of canceling out some of the self-defeating habits of the current economic system and replacing them with effective incentives toward the habits industrial societies need to establish. Many of the forces pushing us toward collapse are a product of the economic mistakes already discussed in this book and in the way these mistakes are reflected in public policy. Those could conceivably be changed in time to matter.

The transition from the economy of abundance that defined the recent past to the economy of scarcity that will define the near future bids fair to open a window of opportunity for change — though this in itself does not guarantee that the changes in question will be for the better. The accumulated burdens of past mistakes weigh heavily enough on the future that changes of this sort won't stave off a great deal of trouble and suffering, but it's possible that a shift to saner policies backed by more realistic economic ideas could cushion the end of the age of abundance, and make it easier to allocate resources to projects that will actually do some good, instead of pursuing policies which — like nearly all the economic policies currently in place in the industrial world — will simply make matters worse. In the following pages I propose to suggest some of the possibilities still open to the industrial world.

Remembering What Worked

At the head of the list of useful steps belongs the recognition that governments might find it useful to stop making the predicament of the industrial world worse than it already is. The decision of the US government to deal with the immediate impacts of the 2008 real estate crash by what is euphemistically called "qualitative easing," and is better known as printing money, is a case in point. Nearly every penny of the money voted by Congress and conjured into being by the Federal Reserve was given to the same huge banks and financial corporations that caused the real estate bubble in the first place. Some of that money went into bailouts of financial corporations; some of it was used to subsidize new and refinanced mortgages for consumers, in the hope that these could be used to prop up one set of unpayable debts with another. The result in both cases has been the perpetuation of the same disastrous imbalances that helped spawn the housing bubble in the first place, while flooding the tertiary economy with even higher levels of unpayable debt.

Such antics have given additional fuel to those who claim that government has no business taking actions affecting the market at all. Still, the fact that incompetent doctors maim and kill their patients on occasion does not mean that all medical care is a bad idea. Markets, despite their benefits, are not uniquely gifted with a perfect ability to make economic decisions. This has become a controversial point in recent years. Just as a great many people on the left have picked up the dubious habit of using labels such as "fascism" for any political system to the right of Hillary Clinton, a great many people on the right seem to have convinced themselves that any form of economic regulation at all is tantamount to some sort of neo-Marxist hobgoblin. It seems to have been forgotten that a spectrum consists of something other than its two endpoints.

Drowning, however, is not the only alternative to dying of dehydration; there's a middle ground that is noticeably more pleasant than either. The same principle applies in economics. The experiment of having government own all the means of production in an industrial society, along the lines proposed by Marx, received a thorough test at the hands of the Communist bloc and failed abjectly to produce the benefits Marx claimed it would produce. At the same time, the experiment of having government keep its hands off the economy altogether in an industrial society, along the lines proposed by a great many free market proponents these days, received an equally thorough test, and failed just as dismally. In America that test ran from the end of the Civil War into the first decade of the twentieth century, and the result was a catastrophic sequence of booms and busts, the transfer of most of the nation's wealth to a tiny minority of wealthy people, the bitter impoverishment of nearly everyone else and a level of social unrest that included two presidential assassinations and so many bomb attacks on the rich and their families that bomb-throwing anarchists became a regular theme of music hall songs.

It's always possible for theorists to contrast a Utopian portrait of a free-market economy against the gritty and unwelcome realities of extreme socialism, just as it's possible for people on the other side of the spectrum to contrast a Utopian portrait of a socialist economy against the equally gritty and unwelcome realities of unfettered capitalism. Both make great rhetorical strategies, since the human mind is easily misled by binary logic: if A is evil, it seems wholly reasonable to claim that the opposite of A must be good. The real world does not work that way, but this is hardly the only case in which rhetoric ignores reality.

Thus it simply isn't true that the best, or for that matter the only, alternative to the unchecked rule of corporate robber barons is Marxist-style state ownership of the economy; once again, dying from heatstroke is not the only alternative to dying from hypothermia. It means, rather, that something between these two extremes might be worth trying, especially if it can be shown by historical evidence to work tolerably well in practice. Of course this is what history shows; broadly speaking, economies that leave the means of production in private hands, but use appropriate regulation to harness their energies to the public good, consistently produce more prosperity for more people than either unfettered capitalism or extreme socialism.

This being said, the midpoint between these extremes may not lie where today's conventional wisdom tends to place it. Consider an example from the not-too-distant past: a large industrial nation with a capitalist economy that had remarkably tough regulations restricting the growth of private fortunes and the abuses to which capitalist economies are so often prone. The wealthiest people in that nation paid more than 90 percent of their annual income in tax, and monopolistic practices on the part of corporations faced harsh and frequently applied judicial penalties. The financial sector was particularly tightly leashed: interest rates on savings accounts were fixed by the government, usury laws put very low caps on interest rates for loans and legal barriers prevented banks from expanding out of local markets or crossing the firewall between consumer banking and the riskier world of corporate investment. Consumer credit was so difficult to get, as a result, that most people did without it most of the time, using layaway plans and Christmas Club savings programs to afford large purchases.

According to the standard rhetoric of free market proponents these days, so rigidly controlled an economy ought by definition to be hopelessly stagnant and unproductive. This shows once again the separation of rhetoric from reality, however, for the nation I have just described was the United States during the presidency of Dwight D. Eisenhower — that is, during one of the most sustained periods of prosperity, innovation, economic development and international influence this nation has ever seen. Now of course there were many other factors behind America's 1950s success, just as there were other factors behind the decline since then; still, it's worth noting that as the economic regulations of the 1950s have been dismantled — in every case, under the pretext of boosting American prosperity — the prosperity of most Americans has gone down, not up.

The economic regulations that shaped the American economy in the 1950s had a curious dimension, though. Outside of antitrust legislation, very little of it applied to the economy of goods and services on any level, whether that of Mom and Pop grocery stores or big industrial conglomerates. The bulk of it, and very nearly all the strictest elements of it, focused on the financial industry. In the terms used in this book, instead of regulating the production and consumption of goods and services in the secondary economy, the economic policies of the Eisenhower era focused on regulating the tertiary economy, ensuring that too much tertiary wealth did not end up concentrated unproductively in too few hands, and controlling its propensity to multiply as enthusiastically as rabbits on Viagra. That distinction between a relatively unregulated secondary economy and a tightly controlled tertiary economy seems to have worked fairly well, just as the removal of all limits on the tertiary economy has consistently ended very badly.

It makes a good measure of how far the United States has come as a nation that the economic policies of one of the most successful twentieth-century Republican administrations would be rejected by most of today's Democrats as too far to the left. A case could be made, in fact, that far and away the most sensible thing the US Congress could do today, in the face of an economy that has very nearly choked to death on its own bubbles, is to reenact the economic legislation in place in the 1950s, line for line. Such proposals outrage the common conviction that going backwards is a bad thing. Still, when you're hiking in the woods and discover that you've taken a trail that leads someplace you don't want to go, your best bet is normally to turn around and go back to the last place where you were still going in the right direction.

Defending the Commons

This strategy of returning to ways of doing things that have worked in the past, however, will not address the conflict between the limitless tertiary economy and the hard limits of the primary economy, which is the ultimate driving force behind the crisis in industrial society. As we've seen, resource depletion imposes costs, and often drastic ones, on the economy. Under current economic arrangements, however, those costs are not charged to the people who benefit from the activities in question. The owner of an oil well, for example, gets the economic benefits of pumping oil out of the ground, but does not have to pay for the impact that today's extraction will have on tomorrow's economy. For many years, in fact, government policies in many industrial nations have actually rewarded oil well owners for accelerating the depletion of the resource and imposing massive costs on the future.

In the same way, the owner of a smokestack that dumps pollution into the atmosphere gets the economic benefits of whatever activity produces the pollution, but does not have to pay for any of the costs incurred throughout the economy as a result of the pollution. This asymmetry has at least two results. First and most obviously, neither the oil well owner nor the smokestack owner has any economic incentive to decrease the negative impacts of his or

her activities. Second, the long-term economic burdens of depletion and pollution are not included in measures of the costs of the
well or the smokestack.

The result is a massive distortion in our understanding of the
realities that shape our lives. It's generally not considered a viable
business plan — outside of today's financial industry, that is — to
make large profits in the short term by running up debts so large
the business will have to declare bankruptcy in the not-too-distant
future. Yet this is exactly what an economic system that ignores
the cumulative costs of resource depletion and pollution is doing,
on an even larger scale. The future costs of extracting resources
from depleted reserves and mitigating the impacts of a polluted
environment have the same effect as the future costs of debt service
on excessive borrowing; they make the act of buying temporary
prosperity in the near future a cause of impoverishment or collapse
further down the road.

Garret Hardin's "The Tragedy of the Commons," as already
discussed, addressed this issue.[2] Hardin showed that when the
benefits from exploiting a resource went to individuals but the
costs were spread throughout the community, individuals intent
on maximizing their own individual benefit would overexploit the
resource and cause the whole community to suffer drastic losses
in the longer run. His logic was impeccable, and there are plenty
of real-world examples of resource exhaustion driven by this very
process, but it has been pointed out by his critics that resources
held in common have in fact been managed sustainably in countless cases around the world and throughout history. The question
that has to be asked is where the difference comes in.

This is where the divide pointed up earlier — the gap between
economic realities and the models our society uses to understand
them and predict their effects — comes into play. Hardin was quite
correct that when individuals got the benefits of resource exploitation without paying their fair share of the costs to the community,

exhaustion of the resource follows. Those societies that have managed their commons successfully, in turn, found ways to make those who gained the benefits of resource exploitation pay a commensurate share of the costs. The collective understanding of economics in these societies, in other words, and the social policies that they adopted to shape economic behavior, took the tragedy of the commons into account and adjusted the customs and laws governing economic exchanges accordingly.

Several thoughtful proposals have been circulated over the years to match up costs and benefits.[3] None of these proposals are likely to be popular any time soon, since the average voter in the industrial world receives substantial benefits from the maltreatment of the primary economy of Nature while avoiding most of the costs; to name only one example, the benefits of automobile transportation and central heating do not come with a bill for the air pollution these activities produce. Still, costs for other forms of commons are already shared by most people in the industrial world on a pay-as-you-go basis; power and sewer bills come to mind as examples, and the same principle may be worth proposing to deal with the costs of maintaining natural commons as well.

The acronym TANSTAAFL — short for "There Ain't No Such Thing As A Free Lunch" — popularized by Robert Heinlein in his science fiction novel *The Moon Is A Harsh Mistress* — makes a good slogan for this approach. The entire industrial world has been treating the primary economy of Nature as a source of free lunches for the last three centuries or so, with increasingly problematic results. One of the reasons why it's important to treat the natural world as an economic force in its own right — the primary economy, in the terminology used in this book — is precisely the need to begin reckoning up harm to natural systems as an economic cost that has to be dealt with in analyses of costs and benefits, not an "externality" that can be swept under the rug by a display of economic legerdemain. As people begin to get used to thinking of trees, bees and mineral cycles as part of the economy that supports them and forms the unavoidable foundation for any real personal or national prosperity, the habit of treating Nature as a disposable amenity will likely begin to give way to more rational and pragmatic ways of thinking.

This will also require, of course, a reorientation of economic statistics. Today's figures, as discussed in detail earlier, ignore the primary economy completely, measure the secondary economy purely in terms of the tertiary — calculating production in dollars, say, rather than potatoes and haircuts — and focus obsessively on the tertiary. This fixation means that if an economic policy boosts the tertiary economy it looks like a good thing, even if that policy harms the secondary or the primary economies, as it very often does. Thus the choice of statistics isn't a neutral factor, or a simple one; if that choice echoes inaccurate presuppositions — for example, the fantasy that the human economy is independent of Nature — it can feed those presuppositions right back in as a distorting factor into every economic decision.

The solution is clear enough: sort out primary, secondary and tertiary inputs in the most important economic statistics. The gross domestic product forms an excellent example. Recalculating this according to the three economies just sketched out, the statisticians of some imaginary Bureau of Honest Figures might produce something like this:

The gross primary product or GPP would be the value of all unprocessed natural products at the moment they enter the economy — oil as it reaches the wellhead, coal as it leaves the mine, grain as it tumbles into the silo and so on — minus all the costs incurred in drilling, mining, growing and so on. (Those belong to the secondary economy.)

The gross secondary product or GSP might be the value of all goods and services in the economy, except for raw materials from Nature and financial goods and services.

The gross tertiary product or GTP might be the value of all financial goods and services, and all money or money equivalents, produced by the economy.

The value of having these three numbers, alongside or instead of one gross domestic product, is that they can be compared to one another and their shifts relative to one another can be tracked. If the GTP balloons and the other two products stay flat or decline, for example, that doesn't mean the country is getting wealthier; it means that the tertiary economy is inflating, and needs to have some air let out of it before it bursts in a speculative crash. If the GSP increases while the GPP stays flat, and the cost of extracting natural resources isn't soaring out of sight, then the economy is becoming more efficient at using natural resources, in which case the politicians and executives have good reason to preen themselves in public. More generally, dividing up the GDP into these components would make it much easier to differentiate between movements in the real economies of goods and services, on the one hand, and the mere multiplication of abstract paper wealth on the other.

Taxing The Right Things

Another tool that deserves more attention than it has received is the seemingly unpromising field of tax policy. It's a subject on which a great deal of nonsense abounds just now. A recent book titled *The End of Prosperity*, for example, claims that if the US government raised taxes to a level that would actually pay for the services it provides, the result would be, as the title claims, the end of prosperity.[4] Somehow the authors managed to ignore the fact that in the 1950s, when American prosperity was by most measures at its all-time peak, people in the upper tax brackets paid more than 90 percent of their income to Uncle Sam, and that the country has by most measures become less prosperous, not more, as those tax rates have been lowered.

There's a reason for that, and it ties back into the distinctions among the primary, secondary and tertiary economies central to the argument of this book. The primary economy of Nature and the secondary economy of the production of goods and services by human labor are natural systems subject to negative feedback loops that tend to hold them in balance. The tertiary economy of money and other forms of abstract wealth, by contrast, is subject to positive feedback loops that drive it out of balance and tend to unbalance the other two economies as well.

The feedback loops of the tertiary economy that cause the most damage are the ones that make accumulations of paper wealth increase exponentially in value. It's thus not hard to recognize that anything that drains off those accumulations—for example, by putting them in the pockets of people who spend their money on groceries instead of more financial paper—tends to stabilize the economy. That's what the high tax rates of the fifties did, and it's thus hardly accidental that the more drastically tax rates have been cut in the last three decades, and the more tertiary wealth has been shielded from taxes by special capital gains tax rates and the like, the more drastically the tertiary economy has manifested its normal tendency to run to extremes and blow itself up in the process.

Still, there are arguably more tightly focused ways to drain off excesses from the tertiary economy, and the existing tax code also meshes very poorly with the primary economy and its emerging reality of scarce resources and overburdened natural cycles. In a world where the accelerating exploitation of natural resources and the accumulation of paper wealth are major sources of problems, while the human labor at the core of the secondary economy is the most renewable resource we have, we arguably tax the wrong things.

Imagine, then, a tax code that taxes the right things instead. In this imaginary code there are no sales taxes and no taxes on income from wages, salaries, dividends, royalties and rents—that is, no taxes on the secondary economy at all. Instead, there are two classes of taxation. The first, on the primary economy, taxes natural goods and services; the second, on the tertiary economy, taxes interest income, capital gains, and all other money made by money.

The taxes on natural goods and services would follow the same rough line of logic as property taxes do at present. Here in the United States, for example, the federal government, as trustee for the American people, effectively owns all the real estate within its borders — when you buy property, what you're buying is the right to use that property within the limits of the laws and the national interest, which is why China can't just contract with private landowners to buy a couple of disused fishing harbors on our west coast as bases for its navy. The same principle could just as reasonably apply to every other resource in the country. When you pump oil from the ground you're depleting part of the patrimony of the nation, and you should have to pay the government for that privilege. When you dump smoke out of a tailpipe, equally, you're using the nation's atmosphere as a kind of gaseous landfill, and you should have to pay to do that.

Every natural resource of every kind, under this system, would thus be subject to an extraction or a pollution tax. Resources mined in other countries and imported, whether in raw form or in finished products, would be subject to an equivalent tariff to prevent manufacturers from stripping the resources of other countries in an attempt to avoid these taxes. The taxes would apply at the point where raw materials from the primary economy of Nature enter the secondary economy, so that most of them would be paid by manufacturers, with the costs passed on to consumers in the form of higher prices.

Now of course this would mean that many goods and services would cost considerably more than they do now. Since everyone would have the money that now goes to income and sales taxes, this may be a little less of a burden, but it's also relevant that people can avoid resource taxes by changing their buying habits. If you buy a hybrid car, you're going to pay a lot less in petroleum tax, and a lot less in tailpipe tax as well — though the extraction taxes folded into the sticker price for the rare earth minerals in the batteries and electronics may set you back a bit, as they should. If you don't own a car at all, you laugh all the way to the bank. Similarly, the price of a product made from metal mined from the earth includes the extraction tax for the ore, but the price of a product made from recycled material doesn't — recycled material is already in the secondary economy and is thus exempt from extraction taxes. This gives manufacturers a big incentive to use recycled material and undercut the competition.

The second set of new taxes targets a different problem. Right now, with the tax laws we have it's to the economic advantage of businesses to pull their money out of producing goods and services and put it into blowing bubbles in the tertiary economy of paper wealth. That's part of the reason why General Motors manufactured more financial paperwork than cars for quite a few years, until it got into a head-on collision with bubble economics. From a broader perspective, that's part of the reason why America produces so few goods and nonfinancial services nowadays, and so much in the way of abstract paper wealth. Taxing financial income, but not earned income, does a fair amount to reverse that equation. If putting your money into bonds or derivatives means that any profit you make suffers a large tax bite, while putting your money into producing goods and services means you pocket the profits tax free, derivatives and bonds will look much less inviting.

Notice also how these two kinds of taxes work to take an even larger bite out of one of the most mistaken economic priorities of our time, the replacement of human labor by machines. Most industrial nations have a serious unemployment problem; even before the most recent financial crisis, good working-class jobs were very hard to come by. There are plenty of reasons why that happened, but tax policies that makes employers pay half again or more of the cost of a worker's wages in order to hire that worker certainly haven't helped. Eliminate those taxes and place taxes on energy and natural raw materials instead, and in a good many cases a worker instead of a machine becomes the most cost-effective way to do the job.

Other arrangements could easily be devised to accomplish the same ends. The point I want to make with this extended example, though, is that some of the ways we do business, and pay (or don't pay) for government services, don't fit the realities of an age of ecological limits. A tax code that burdens the secondary economy while encouraging the wasteful plundering of Nature and the bubble-blowing antics of the tertiary economy, is not going to help us weather the storms of the near future. Any tax code that does the opposite—that makes it more profitable to employ human labor to meet human needs, and less profitable to disrupt the natural cycles that undergird our survival or to feed speculative excesses that pump imbalances into an already troubled economy—could be a very helpful asset in a time of crisis, and could be put in place tolerably easily, without having to tear an entire society to pieces and rebuild it from the ground up.

Would such a new tax code solve all our problems? Of course not. Still, the attitude that we have to find a single solution that deals with all our problems is a large reason why constructive responses to those problems seem so far out of reach. If we're to face the difficult future ahead of us with any sort of grace, that achievement is much more likely to happen as a result of careful, tentative steps than it is to unfold from some grandiose plan to reach a perfect world in a single leap. Monkeying with the tax code so that it rewards the economic behaviors that might help us get through the approaching troubles, rather than rewarding the economic behaviors that will only make things worse, is only one step in the necessary direction—but it's one step further than any industrial nation has yet gone, and thus is arguably worth taking.

Housebreaking the Corporations

Another set of changes—mutations, to return to the biological metaphor—worth considering here involves changing the way that business corporations relate to the public good. This subject has seen a great deal of discussion in the alternative press over the last few years, much of it focused on the belief that corporations are something close to evil incarnate. The entire topic could use an approach less burdened with mythic archetypes. It is true, of course, especially in America, that corporate misbehavior has become common, and corporations have come to have an inappropriate degree of influence on the political process, but this latter could be said with equal truth of nearly any other large and well-funded institution in contemporary life, from the retiree lobby to the American Medical Association.

The source of the political power of corporations, and that of other influential groups, deserves to be studied carefully. Despite the rhetoric of impending tyranny retailed equally by the far left and the far right these days, political power in most of today's industrial nations is not particularly centralized; rather, it's highly diffused, distributed among the quarrelsome factions of a large and fractious political class. No one faction can carry out its agenda without the consent of many others, and the result is that any well-organized faction becomes a power broker. It can drive whatever bargains it wishes in exchange for supporting the agendas of other factions, and use that clout to defend those positions it considers nonnegotiable.

A good deal of the stalemate that puts necessary reforms out of reach in most of the capitals of the industrial world just now is a function of this process. It's hard to think of any such reform that won't step on the toes of at least one well-funded faction, and so business as usual proceeds on its merry way, even though almost everyone recognizes that the end result will be to nobody's benefit. Consider, as one example out of many, the way that the retiree lobby keeps the simplest and sanest response to America's looming budget crisis — a means test on Social Security and Medicare — from even coming up for discussion. The same process, enforced by one or another power center within the political class, puts every other effective reform equally out of reach.

Corporations are merely one set of power centers making use of this system to pursue their own advantage at the cost of the general welfare. Certain details of the relationship between corporations and society, though, make the conflict between corporate interests and public welfare particularly frequent. Corporations, under the laws of the United States and most other nations, are legal persons; they have many, though not all, of the same rights that "natural persons" — that is, you and me — have under the law. The most obvious difference these days between corporations persons and the other power centers that influence our collective decisions is that the corporate kind are noticeably more antisocial. They pursue their purposes — the production of tertiary wealth — with a single-mindedness and a lack of concern for consequences that, in natural persons, would be accurately labeled psychopathic. They've proven themselves consistently willing to lie, cheat, steal and kill whenever the likely return on these acts outweighs the risk of punishment.

E. F. Schumacher, who was amply familiar with these habits of corporate misbehavior, hoped to solve them by encouraging different models of corporate governance. One of his central examples was a British firm, Scott Bader & Co. Ltd., which was organized as a partnership of its own employees.[5] Similar schemes, ranging from cooperatives to employee-owned corporations, have been put to the test over the years, and work at least as well as the usual style of corporate management. Still, no one has yet found a way to convince stockholders and corporate executives to voluntarily accept a reorganization that would sharply decrease their wealth and influence, and until some such means can be found, Schumacher's approach seems unlikely to make a large difference.

Some writers in recent years, in response to the parade of corporate abuses, have simply turned up the rhetoric of moral denunciation, with the usual results — a warm glow of self-righteousness on the part of the denouncer and no effect at all on the denouncee. Others have proposed various means for forcing corporate persons to behave themselves. One popular proposal would subject corporations to regular review by some independent body which could annul the charter of any corporation that refused to be properly housebroken, forcing its dissolution.[6] What would keep this body honest in the face of the fantastic potential for corruption, the proponent of this notion does not say, but there's another issue here. We've actually got a tolerably effective way of responding to antisocial behavior; we just don't apply it to corporate persons the way we do to natural persons.

A glance back into the history of law may help clarify the matter. I'm not sure how many people these days know that the law codes of most European nations in the early Middle Ages operated almost entirely on the basis of fines. The principle of wergild, as it was called, gave each person a cash value; if a murder took place, the murderer had to pay the family of the victim that cash value as wergild for the death. Lesser injuries and insults called for lesser fees. It didn't work very well, not least because anyone who had enough money could act the way mainstream economists think we all ought to act, on the basis of a simple cost-benefit analysis. If you could afford to pay the wergild for killing somebody and decided it was worth the expense, why not?

So as the Dark Ages gave way to less chaotic times, legal codes across Europe replaced wergild with punishments that were a good deal less easy to shrug off. This is why natural persons who

are convicted of felonies, by and large, can't get away with just paying a fine; they go to jail, or if the crime is heinous enough and it
happens in a jurisdiction with capital punishment, they die. While
this approach also has its failings, it does seem to deter criminal
activity a good deal more effectively than the wergild principle
once did.

From this perspective, the problem with corporate persons is
simple enough. The only risk they run in breaking the law is that
they have to pay wergild, and that doesn't constrain antisocial behavior any more effectively now than it did in the Dark Ages.

My more perceptive readers may be wondering at this point
whether I'm seriously proposing that corporations should be
thrown in jail or put to death. This is indeed what I'm proposing,
with the adjustments needed to account for the differences between corporate persons and natural persons. The essential nature
of imprisonment for a crime, after all, is that the criminal ceases
to be a free person. For a specified period of time he is a chattel of
society, and society has the right to profit from his labor during
that period. And capital punishment? The criminal, having proven
that he isn't willing to abide by even the most minimal standards of
social existence, ceases to exist by act of society. Both of these can
be applied to corporations easily enough.

Imagine, then, that a corporation—we'll call it the Shyster
Company—has just been caught deliberately selling worthless securities to widows and orphans. The district attorney files charges
of felony fraud and theft in state court. The trial date arrives, the
lawyers bicker, the jury finds the defendant guilty as charged and
the judge sentences the corporation to ten years in the slammer. In
practice, what happens is that the judge appoints a trustee, who
takes control of Shyster and all its assets. For the next ten years,
Shyster is a wholly owned subsidiary of the state government. Its
stock pays no dividends and has no voting rights, its directors have
to find something else to do with their time, and if the trustee

decides that the CEO and other overpaid office fauna have to find new jobs, they have to find new jobs — assuming that they're not doing time themselves, as they very well could be. All profits earned by Shyster during its period of imprisonment go to the state government, subject to set-asides that pay restitution to the victims of the crime.

Meanwhile another conglomerate — we'll call this one Dirty Rotten Scoundrel Inc. — has been caught knowingly selling food products tainted with deadly bacteria, and a dozen people have died. This time the district attorney files charges of aggravated first degree murder. The trial date arrives, the media has a field day, the lawyers bicker, the jury returns a verdict of guilty as charged, the judge sentences DRSI to death and the appeals court upholds the sentence. In practice, what happens is that on the scheduled date of execution, DRSI ceases to exist. Its stock becomes worthless, its assets are sold off in an auction in which no former shareholder is allowed to bid, its name and trademarks can never again be used by anybody under penalty of law and its creditors get whatever scraps are left once the victims' families receive their settlements.

It's crucial that the stockholders in both cases, and the creditors in the latter case, suffer for the behavior of the corporation. The stockholders of a corporation are its owners in fact and law; they profit from its activities and therefore should pay for its crimes. The laws governing corporations limit the liability of stockholders to the value of their investment, and there's no need to overturn that principle; it simply needs to be applied to criminal cases in the same way that responsibility is applied in cases involving natural persons. Equally, those who loan money to a business already accept the risk that the business may go bankrupt and their money may be lost; making corporations subject to criminal penalties simply applies the same principle of risk to the possibility that a corporation may violate the law and pay for its crimes.

Thus, under this system if word gets out that a corporation is pushing the limits of legality, the stockholders have a very strong incentive to sell, driving down the value of the stock. Equally, if lenders become aware that a corporation is engaging in really egregious behavior, they have a very strong incentive to charge higher interest rates or even to stop loaning money to the corporation. Neither has any such incentive under the current system, which is one reason why corporations act as though their quarterly profit statements are the only things that matter. To their stockholders and creditors, this is essentially the case; this proposal would change that.

Finally, governments have a powerful incentive to enforce the law, which current corporate regulation schemes generally lack. Governments always need money; raising taxes is unpopular, but catching a crooked corporation that has violated the law and making the rascals pay for their crime makes excellent press, and five or ten years of corporate income in the state treasury would gladden the heart of even the most unregenerate corporate stooge in the state legislature.

Readers may be wondering at this point whether I think such a proposal has the chance of a snowball in Beelzebub's back yard of being enacted. As it happens, I do. One of the repeated lessons of history is that the political power of business waxes during times of relative stability and crumples in times of turmoil and crisis. The long European peace of the nineteenth century saw business interests dominate most Western governments to an extent that would be considered extreme even by today's lax standards; when that peace shattered in 1914 it took the power of big business with it, and by the time the rubble stopped bouncing in 1954, every Western country had either embraced some degree of socialism outright, or adopted radical economic reforms that would have been considered unthinkable before a flurry of bullets at Sarajevo tipped the world into chaos.

We are facing a similar age of crisis now, and the wide diffusion of power common in ages of prosperity is usually an early casualty of such times. As current distributions of power break down and today's veto groups find themselves increasingly shut out of the emergency coalitions and charismatic governments that tend to rise in difficult eras, much of what passes for corporate power will inevitably turn out to be much more feeble than it looks today. Several countries, with Russia in the lead, have already seen showdowns between corporate influence and the power of national governments, and the corporations have not fared well in these battles. When the power of money faces off against the power of violence, money comes out a distant second.

As the Great Recession deepens, the decline of world petroleum production begins to bite and rising world powers contend with declining America and each other to settle whose will be the next global empire, the superiority of political and military force to money's influence is likely to play an increasingly large role in determining the balance of power. I suspect that by the time the current mess gets much deeper, business interests will be facing organized efforts to do things much more drastic than simply holding corporations responsible for their crimes, and may be willing to bargain in the hope of survival in exactly the same way their predecessors did in that earlier time of troubles.

A Crisis of Complexity

Another set of ideas commonly suggested in response to the crisis of industrial society are more problematic than the reforms already suggested. These are the programs that seek to deal with the problems of our time by placing a flurry of additional laws and regulations, and additional bureaucracies to enforce them, on top of the existing laws, regulations and bureaucracy we have now. Choose a crisis, from the puncturing of the housing bubble in 2008 to the 2010 Deepwater Horizon oil spill, and you can count on finding someone — more often, quite a few someones — who think the best way to respond to it is to impose more regulations, hire more bureaucrats and provide more government grants. On the plane of popular culture, the same habit of thought pervades the often-repeated suggestion that the crisis du jour can be solved by a Manhattan Project or Apollo program, government-run and lavishly funded with tax dollars.

I am far from certain that this is a good idea. As Joseph Tainter has pointed out in his valuable book, *The Collapse of Complex Societies*, increases in social complexity are subject to the same law of diminishing returns as anything else, and sooner or later a society that responds to every challenge by adding a new layer of complexity will reach the point where adding more complexity causes more problems than it solves. Today's industrial societies, pushing the curve of diminishing returns with the help of abundant fossil fuel energy, have become more complex than any other human society in recorded history. Several observations concerning Tainter's insight are thus worth making here.

First, the diminishing returns of complexity do not strike everywhere at once but appear here and there at first, most often in whatever institutions or areas of life have been most heavily loaded with excess complexity. Thus, a society that has overloaded itself with complexity will tend to heap up more complexity in some areas of life than others, and one or more of these areas will normally tip over into dysfunction sooner than others. Thus a society that is hammered by repeated crises of the same kind, and tries to solve them with layers of additional complexity that consistently seem to make the problem worse, may be at risk of tipping over into a wider dysfunction of which the visible crises are merely symptomatic.

Second, if a society has driven itself past the point of negative returns on complexity, and continues to try to add complexity to solve the resulting problems, it risks establishing a disastrous feedback loop in which its attempts to solve its problems become the major source of new problems. This can also apply to specifics as well as generalities, and show up first in particular aspects of a society's collective life.

Third, one of the ironies faced by a society that has reached the point of negative returns on complexity as a means of problem solving is that thereafter, the only way it can solve its problems is by not solving its problems. Any attempt to solve problems by adding additional complexity will simply make matters worse, while allowing some particularly problematic heap of complexity to crash and burn may just reduce the complexity of the whole system to a point at which something constructive can actually be done. In the extreme case, where an entire society has pushed itself past the point of negative returns on complexity, political and economic collapse can be an adaptive response to a rising spiral of crisis that can be ended in no other way.

Now the logical solution to a problem caused by too much complexity is to reduce the amount of complexity. Tainter argues that societal collapse has exactly this function; when a society has backed itself into a corner by heaping up more complexity than it can manage, collapse offers the only way out. Still, a proposal to reduce the complexity of contemporary civilization can count on getting no interest from the political classes of today's industrial nations, or for that matter from the population at large. The experiment has been tried, after all; it's worth remembering the extent to which the baby steps toward lower complexity taken in the 1970s helped to fuel the Reagan/Thatcher backlash of the 1980s.

The hope of a mass conversion to sustainability by political means — by legislation, let's say, backed up by the massive new bureaucracy that would be needed to enforce "green laws" affecting every detail of daily life — is yet another attempt to solve a complexity-driven problem by adding on more complexity. That's a popular strategy these days. It makes sense to most of us since it's the sort of thing we're used to doing, and it would provide a large number of economic and social niches for specialists — in this case, members of the professional activist community, who might reasonably expect to step into staff positions managing the new level of complexity.

All of this is very familiar ground, echoing as it does the way that countless other efforts at reform have turned into layers of complexity in the past. To suggest, as I do, that it won't work, doesn't mean that it won't be tried. Quite the contrary — as the industrial age winds down, very nearly every plausible attempt to solve the problems of complexity with more complexity will likely get at least some funding, and be given at least a token trial.

Much the same process is already under way in the energy field. We've already had the corn ethanol boom here in the US; the cellulosic ethanol and algal biodiesel booms have been delayed by the impact of a collapsing economy on credit markets, but somebody will doubtless find a way around that in good time. Down the road a bit, a crash program to build nuclear power plants is pretty much a foregone conclusion; fusion researchers will have the opportunity to flush billions more dollars down the same rat hole they've been attempting to fill since the 1950s. You name it, if it's complex and expensive it will get funding.

Not all of that money will be entirely wasted, either. Current wind power technologies and PV panels may not be sustainable over the long term, but for the decades immediately ahead they're an excellent investment; anything that can keep the grid supplied with power as fossil fuel production drops out from under the world's industrial economies may be able to help make the next few decades less brutal than they might otherwise be. With any luck, there'll be a boom in home insulation and weatherstripping, a boom in solar hot water heaters, a boom in backyard victory gardens and the like — small booms, probably, since they aren't complex and expensive enough to catch the contemporary imagination, but even a small boom might help.

On the whole, though, the pursuit of complexity as a solution for the problems caused by complexity is a self-defeating strategy. It happens to be the self-defeating strategy to which we're committed, collectively and in most cases individually as well, and it can be dizzyingly hard for many people to think of any action at all that doesn't follow it. Can you think of a way to deal with the problems of complexity in today's industrial nations — problems that include, but are not limited to, rapidly depleting energy supplies, ecological destruction and accelerating economic turbulence — that doesn't simply add another layer of complexity to the mess?

There is at least one such way, pioneered by Schumacher in a different context. It starts with what he termed the Principle of Subsidiary Function.[7] This rule holds that the most effective arrangement to perform any function whatsoever will always assign that function to the smallest and most local unit that can actually perform it.

It's hard to think of any principle that flies more forcefully in the face of every presupposition of the modern world. Economies of scale and centralization of control are so heavily and unthinkingly valued that it rarely occurs to anyone that in many situations they might not actually be economies at all. Still, Schumacher was not a pie-in-the-sky theorist; he drew his conclusions on the basis of most of a lifetime as a working economist in the business world. He noticed that the bigger and more centralized an economic or political system happened to be, the less effectively it could respond to the complex texture of local needs and possibilities that makes up the real world.

This rule can be applied to any aspect of the predicament of industrial society you care to name. Attempts to make such a response on the highest and least local level possible – for example, the failed international negotiations to establish a global response to anthropogenic climate change — have done quite a respectable job of offering evidence for Schumacher's contention. Attempts to do the same thing at a national level aren't doing much better. The lower down the ladder you go, and the closer you get to individuals and families confronting the challenges of their own lives, the more success stories you find.

By the same logic, the best place to start backing away from an overload of complexity is in the daily life of the individual. What sustains today's social complexity, in the final analysis, is the extent to which individuals turn to complex systems to deal with their needs and wants. To turn away from complex systems on that individual level, in turn, is to undercut the basis for social complexity, and to begin building frameworks for meeting human needs and wants of a much simpler and thus more sustainable kind.

It also has the advantage — not a small one — that it's unnecessary to wait for international treaties or government action or anything else to begin having an effect on the situation. It's possible to begin right here, right now, by identifying the complex systems on which you depend for the fulfillment of your needs and wants and making changes in your own life to shift that dependency onto smaller or more local systems or onto yourself or onto nothing at all — after all, the simplest way to deal with a need or want, when doing so is biologically possible, is to stop needing or wanting it.

Such personal responses have traditionally been decried by those who favor grand collective schemes of one kind or another. Still, it deserves to be remembered that a small step that actually happens will do more good than a grandiose plan that never gets off the drawing board, a fate suffered by nearly all of the last half century's worth of grandiose plans for sustainability. Starting from personal choices and local possibilities, rather than abstract and global considerations, makes it a good deal more likely that whatever evolves out of the process might actually work. Tackling the crisis of industrial society from the top down has been tried over and over again by activists for decades now with no noticeable results. Maybe it's time to try something else.

Back To The Future

One application of the principle of subsidiary function that has already received serious attention within the peak oil community is the need to relocalize economic activity — that is, to shift from centralized production of goods and services and continental or intercontinental supply and distribution chains to regional or local production of goods and services using locally available resources. Relocalization is a standard event in ages of contraction. When complex societies overshoot their resource bases and decline, centralized economic arrangements fall apart, long distance trade declines sharply and the vast majority of what we now call consumer goods get made at home or very close to home. That violates much of the conventional wisdom that governs economic decisions these days: centralized economic arrangements are thought to yield economies of scale that make them more profitable than decentralized local arrangements.

When history conflicts with theory, though, it's not history that's wrong, so a second look at the conventional wisdom is in order. The economies of scale and the resulting profitability of centralized economic arrangements don't happen by themselves. They depend, among other things, on transportation infrastructure. This doesn't happen by itself, either; it happens because governments pay for it, for purposes of their own. The Roman roads that made the tightly integrated Roman economy possible, for example, and the interstate highway system that does the same thing for America, were not produced by entrepreneurs; they were created by central governments for military purposes. (The legislation that launched the interstate system in the US, for example, was pushed by the Department of Defense, which wrestled with transportation bottlenecks all through the Second World War.)

Government programs of this kind subsidize economic centralization. The same thing is true of other requirements for centralization — for example, the maintenance of public order, so that shipments of consumer goods can get from one side of the country to the other without being looted. Governments don't establish police forces and defend their borders for the purpose of allowing businesses to ship goods safely over long distances, but businesses profit mightily from these indirect subsidies nonetheless.

When civilizations come unglued, in turn, all these indirect subsidies for economic centralization go away. Roads are no longer maintained, harbors silt up, bandits infest the countryside, migrant nations invade and carve out chunks of territory for their own and so on. Centralization stops being profitable, because the indirect subsidies that make it profitable aren't there any more.

The decline and fall of the Roman Empire, a well-documented example, was a process of radical relocalization, and the result was the Middle Ages. The Roman Empire handled defense by putting huge linear fortifications along its frontiers; the Middle Ages replaced this with fortifications around every town and baronial hall. The Roman Empire was a political unity where decisions affecting every person within its borders were made by bureaucrats in Rome. Medieval Europe was the antithesis of this, a patchwork of independent feudal kingdoms the size of a Roman province, which were internally divided into self-governing fiefs, those into still smaller fiefs and so on, to the point that a village with a fortified manor house at its center could be an autonomous political unit with its own laws and the right to wage war on its neighbors.

The same process of radical decentralization affected the economy as well. The Roman economy was just as centralized as the Roman polity; in major industries such as pottery, mass production at huge regional factories was the order of the day, and the products were shipped out via sea and land for anything up to a thousand miles to the end user.[8] That came to a halt when roads weren't repaired any more, the Mediterranean became pirate heaven and too many of the end users were getting dispossessed and dismembered by invading Visigoths. The economic system that evolved to fill the void left by Rome's implosion was thus every bit as relocalized as a feudal barony, and for exactly the same reasons.

That system was based on craft guilds, which worked in a distinctive and—to modern minds—highly counterintuitive way.[9] Each city—and "city" in this context means anything down to a town of a few thousand people—was an independent economic center providing skilled trades to a fairly small region. The city might have a few industries of more than local fame, but most of its business consisted of manufacturing and selling things to its own citizens and the surrounding countryside. The manufacturing and selling was managed by guilds, which were cooperatives of master craftsmen. To get into a guild-run profession you had to serve an apprenticeship, usually seven years, during which you got room and board in exchange for learning the craft and working for your master. You then became a journeyman and worked for a master for wages, until you could produce your masterpiece—that's where the word comes from—which was an example of craftwork fine enough to convince the other masters to accept you as an equal. Then you became a master, with voting rights in the guild.

The guild had the responsibility under feudal municipal laws to establish minimum standards for the quality of goods, to regulate working hours and conditions and to control prices. Economic theories of the time held that there was a "just price" for any good or service, usually the price that had been customary in the region since time out of mind, and municipal authorities could be counted on to crack down on attempts to charge more than the just price. Most forms of competition were off limits; if you made your apprentices and journeymen work evenings and weekends to out-produce your competitors, for example, or sold goods below the just price, you'd get in trouble with the guild and could be barred from doing business in the town. The only form of competition that was encouraged was to make and sell a superior product.

This was the secret weapon of the guild economy, and it helped drive an age of technical innovation. As Jean Gimpel showed in *The Medieval Machine*, the stereotype of the Middle Ages as a period of technological stagnation is utterly off the mark. Medieval craftsmen invented the clock, the cannon and the movable-type printing press, perfected the magnetic compass and the water wheel, and made massive improvements in everything from shipbuilding and steelmaking to architecture and windmills, among many other things. The competition between masters and guilds for market share when quality and innovation were the only fields of combat wasn't the only force behind these transformations, to be sure — the medieval monastic system, which put a good fraction of intellectuals of both genders in sheltered workshops where they could use their leisure for any purpose that could be chalked up to the greater glory of God, was also a potent factor — but it certainly played a massive role.

Advocates of relocalization in the age of peak oil may thus find it useful to keep the medieval example in mind while planning for the economics of the future. Relocalized communities must be economically viable or they will soon cease to exist, and while viable local communities will be possible in the future — just as they were in the Middle Ages — the steps that will be necessary to make them viable may require some serious rethinking of the habits that now shape our economic lives.

A Future of Victory Gardens

Plenty of people have argued that the only valid response to the rising spiral of crisis faced by industrial civilization is to build a completely new civilization from the ground up on more idealistic lines. Even if that latter phrase wasn't a guarantee of disaster — if there's one lesson history teaches, it's that human societies are organic growths, and trying to invent one to fit some abstract idea of goodness is as foredoomed as trying to make an ecosystem do what human beings want — we no longer have time for grand schemes of that sort. To shift metaphors, when your ship has already hit the iceberg and the water's coming in, it's a bit late to suggest rebuilding it from the keel up according to some new scheme of naval engineering.

An even larger number of people have argued with equal zeal that the only valid response to the predicament of our time is to save the existing order of things, with whatever modest improvements the person in question happens to fancy, because the alternative is too horrible to contemplate. They might be right, too, if saving the existing order of things was possible, but at this point it's not. A global civilization that is utterly dependent for its survival on ever-expanding supplies of cheap abundant energy and a stable planetary biosphere is not going to make it in a world of ever-contracting supplies of scarce and expensive energy and a planetary biosphere that the civilization's own activities are pushing into radical instability. Again, when your ship has already hit the iceberg and the water's coming in, it's not helpful to insist that the only option is to keep steaming toward a distant port.

What that leaves, to borrow a useful term from one of the most insightful books of the last round of energy crises, is muddling through. Until recently, at least, Warren Johnson's *Muddling Toward Frugality* seemed to have dropped into the limbo our cultural memory reserves for failed prophecies;[10] neither he nor, to be fair to him, anybody else in the sustainability movement of the seventies had any idea that the collective response of most industrial nations to the approach of the limits to growth would turn out to be a 30-year vacation from sanity in which short-term political gimmicks and the wildly extravagant drawdown of irreplaceable resources would be widely mistaken for permanent solutions.

That put paid to Johnson's hope that simple, day-by-day adjustments to dwindling energy and resource supplies would cushion the transition from an economy of abundance to one of frugality.

His strategy, though, still has one thing going for it that no other available approach can match: it can still be applied this late in the game. If it's done with enough enthusiasm or desperation, and with a clear sense of the nature of our predicament, it could still get a fair number of us through the mess ahead; and it certainly offers better odds than sitting on our hands and waiting for the ship to sink, which under one pretense or another is the only other option open to us right now.

A strategy of muddling doesn't lend itself to nice neat check-lists of what to do and what to try, so I won't presume to offer a step-by-step plan. Still, showing one way to muddle, or to begin muddling, and outlining some of the implications of that choice, can bridge the gap between abstraction and action and suggest ways that those who are about to muddle might approach the task — and of course there's always the chance that the example might be applicable to some of the people who read it. With this in mind, I want to talk about victory gardens.

As the rising spiral of economic trouble continues, we can ex-pect drastic volatility in the value and availability of money — and here again, remember that the word "money" refers to any form of wealth that only has value because it can be exchanged for some-thing else. Any economic activity that is solely a means of bringing in money will be held hostage to the vagaries of the tertiary econ-omy, whether those express themselves through inflation, credit collapse or what have you. Any economic activity that produces goods and services directly for the use of the producer and his or her family and community will be much less drastically affected by these vagaries. If you depend on your salary to buy vegetables, for example, how much you can eat depends on the value of money at any given moment; if you grow your own vegetables, using your own kitchen and garden scraps to fertilize the soil and saving your own seed, you have much more direct control over your vegetable supply.

The victory garden as a social response to crisis was an invention of the twentieth century. Much before then, it would have been a waste of time to encourage civilians in time of war to dig up their back yards and put in vegetable gardens because nearly everybody who had a back yard had a kitchen garden in it. That was originally why houses had back yards: the household economy, which produced much of the goods and services used by people in pre-petroleum Europe and America, didn't stop at the four walls of a house. Garden beds, cold frames and henhouses in urban backyards kept pantries full, while no self-respecting farm wife would have done without the garden out back and the dozen or so fruit trees close by the farmhouse.

Those useful habits only went into decline when rail transportation and the commercialization of urban food supplies gave birth to the modern city in the course of the nineteenth century. When 1914 came around and Europe blundered into the carnage of the First World War, the entire system had to be reinvented from scratch in many urban areas, since the transport networks that brought fresh food to the cities in peacetime had other things to do, and importing food from overseas became problematic for all combatants in a time of naval blockades and unrestricted submarine warfare. The lessons learned from that experience became a standard part of military planning thereafter, and when the Second World War came, well-organized victory garden programs shifted into high gear, helping to take the hard edges off food rationing. It's a measure of their success that despite the massive mismatch between Britain's wartime population and its capacity to grow food and the equally massive challenge of getting food imports through a gauntlet of U-boats, food shortages in Britain never reached the level of actual famine.

In the seventies, in turn, the same thing happened on a smaller scale without government action; all over the industrial world, people who were worried about the future started digging victory gardens in their back yards, and books offering advice on backyard gardening became steady sellers. (Some of those are still in print today.) These days, sales figures in the home garden industry reliably jolt upwards whenever the economy turns south or something else sends fears about the future upwards; for many people, planting a victory garden has become a nearly instinctive response to troubled times.

It's fashionable in some circles to dismiss this sort of thing as an irrelevance, but such analyses miss the point of the phenomenon. The reason that the victory garden has become a fixture of our collective response to trouble is that it engages one of the core features of the predicament individuals and families face in the twilight of the industrial age, the disconnection between the tertiary economy on the one hand and the primary and secondary economies on the other. The value you get from a backyard garden differs from the value you get from your job or your savings in a crucial way: money doesn't mediate between your labor and the results. If you save your own seeds, use your own muscles and fertilize the soil with compost you make from kitchen and garden waste—and many gardeners do these things, as a matter of course—the only money your gardening requires of you is whatever you spend on beer after a hard day's work. The vegetables that help feed your family are produced by the primary economy of sun and soil and the secondary economy of sweat; the tertiary economy has been cut out of the loop.

Now it will doubtless be objected that nobody can grow all the food for a family in an ordinary back yard, so the rest of their food remains hostage to the tertiary economy. This is more or less true, but it's less important than it looks. Even in a really thumping depression, very few people have no access to money at all; the problem is much more often one of not having enough money to get everything you need by way of the tertiary economy. An effective response usually involves placing any economic activity that can be done without the involvement of money outside the reach of the tertiary economy and prioritizing whatever money can be had for those uses that require it.

You're not likely to be able to grow field crops in your back yard, for example, but grains, dried beans and the like can be bought in bulk very cheaply at present, and will almost certainly be available in bulk across North America even in times of turmoil. What can't be bought cheaply now, and in a time of financial chaos may not be for sale at all, are exactly the things you can most effectively grow in a backyard garden, the vegetables, vine and shrub fruits, eggs, chicken and rabbit meat, and other foods that provide the vitamins, minerals and nutrients you can't get from 50-pound sacks of rice and beans. These are the sorts of things people a century and a half ago produced in their kitchen gardens, along with medicinal herbs to treat illnesses and maybe a few dye plants for homespun fabric; these are the sorts of things that make sense to grow at home in a world where the economy won't support the kind of abundance most people in the industrial world take for granted today.

It will also doubtless be objected that even if you reduce the amount of money you need for food, you still need money for other things, and so a victory garden isn't an answer. This is true enough, if your definition of an answer requires that it simultaneously solves every aspect of the predicament of industrial society. Waiting for the one perfect answer to come around is a refined version of doing nothing while the water rises. Muddling requires many small adjustments rather than one grand plan: planting a victory garden in the backyard is one adjustment to the impact of a dysfunctional money economy on the far from minor issue of getting food on the table; other impacts will require other adjustments.

A third objection is that not everybody can plant a victory garden in the backyard. This is true enough; a good many people don't have backyards these days, and some of those who do are forbidden by restrictive covenants from using their yards as anything but props for their home's largely imaginary resale value. Still, a victory garden is simply an example of the way that people have muddled through hard times in the past, and might well muddle through the impending round of hard times in the future. If you can't grow a garden in your backyard, see if there's a neighborhood allotment program that will let you garden somewhere else, or look for something else that will let you meet some of your own needs with your own labor without letting money get in the way.

That latter, of course, is the central point of this example. At a time when the tertiary economy is undergoing the first stages of an uncontrolled and challenging simplification, if you can disconnect something you need from that economy, you've insulated a part of your life from at least some of the impacts of the chaotic resolution of the mismatch between limitless paper wealth and the limited real wealth available to our species on this very finite planet.

Now of course it's true that the cost of equipping somebody to perform some economic function locally has already entered the rising collective conversation around peak oil in an informal way. What Transition Towns proponent Rob Hopkins calls "the great reskilling"—the process by which individuals who have no productive skills outside a centralized industrial economy learn how to make and do things on their own—has had to take place within the tolerably strict constraints of what individuals can afford to buy in the way of tools and workspaces, since there isn't exactly a torrent of grant money available for people who want to become blacksmiths, brewers, boat builders or practitioners of other useful crafts.

It may be worth suggesting, though, that Schumacher's theory of intermediate technology might be worth applying directly by those individuals and communities who are willing to put that project into practice. The less it costs in terms of energy and other resources to prepare a community to deal with one or more of its economic needs, after all, the more will be available for other projects. Equally, the more good ideas that can be garnered from the dusty pages of publications issued by Schumacher's Intermediate Technology Development Group and its many equivalents and put to work during the industrial world's decline to Third World status, the more creativity can be spared for other challenges.

After Retirement

The transformations that will force the expansion of victory gardens also promises, as Chapter Three suggested, to bring an end to one of the most basic expectations of people in today's industrial world. Retirement as a social habit was entirely a product of the zenith of the age of abundance. For a brief window of time—rather less than a century—it made financial and political sense for nations in the developed world to pay their elderly citizens to stay out of the work force, in order to keep unemployment down to politically bearable levels. All this unfolded, in turn, from an industrial economy so lavishly supplied with cheap energy that human labor was worth replacing with machines wherever the technology permitted, and so greedy for new markets that every part of human life was made subject to market forces.

Before that period began, something less than half of all economic activity even in the industrial world had anything to do with the market at all. Most women, and many men outside the age of regular employment, worked in a household economy governed by custom and intrafamily exchange rather than market forces. This included essentially everyone who would be eligible for retirement by the standards of the age that has just ended. Outside the market but not outside the demand for skilled human labor, elderly people typically provided household goods and services to a household somewhere in their extended family. That was their full-time job; by contributing the value of their labor and skills, they earned their keep.

The end of the age of cheap energy means that such household economies will again be viable. It also means that they will again be necessary. When the limited energy and resources of a contracting society have to be prioritized for urgent needs, takeout meals and convenience foods will sooner or later draw the short straw; in their absence, most food will once again be made at home from raw ingredients. When the energy cost of the global network of sweatshops that keeps Americans clothed can no longer be met, clothing will once again be made at home from raw fiber, as it was not so long ago, and so on. All this requires human labor. Thus a society no longer supplied with nearly unlimited amounts of cheap abundant energy will have every incentive to keep elderly people in the household labor force, and neither the incentive nor the resources to keep them in comfortable idleness.

Now of course it's true that we will not be landing in such a society overnight. It's also true that the clout of the retiree lobby in most industrial nations is such that public and private pensions will be looted, or simply dissolve in a general financial collapse, only when every other option has been exhausted—though in the United States, at least, the vast tide of red ink currently flooding out of Washington DC is likely to bring about this eventuality sooner rather than later. Still, it's quite possible that at least some of today's retirees and soon-to-be-retirees will manage to cling to that status, at least for a while.

If I were asked for advice about retirement, then, it would probably go something like this: If you're already retired, or within a few years of retirement, it's probably worthwhile to put any investment money you have left into a stable investment, if you can find one. Still, it's unwise to assume that your investments will be worth anything in the long term, and having a Plan B would be a very good idea. If you're more than a decade or so out from retirement, having a Plan B in place is essential. If you're 30 years out or more out, as I am, forget about Plan A for now; you can look into the options for investment later, once the wreckage of the last few decades has been hauled away and a new economic order has begun to take shape, but you probably will never retire.

What sort of Plan B might work best depends on so many local and personal variables that specifics would be misleading. If you've got a large family with whom you're on good terms, bone up on your home economics skills; ten years from now, when four of your grandkids, their spouses and their children live in one rundown McMansion, having Grandma and Grandpa there to cook meals, tend children, and tend the garden will likely be worth much more than your keep. If you don't have a family or can't stand them, cultivate relationships with younger friends, or get ready to take up a second career that you can continue into advanced old age. No matter what you choose, it's not going to be as much fun as sitting on a lawn chair in a Sun Belt trailer park, but then the future is under no obligation to limit itself to those options we prefer.

More generally, the world of money will become much less relevant to most people in the years to come. Few of us will have the option of separating themselves completely from the money economy for many years to come; as long as today's governments continue to function, for example, they will demand money for taxes, and money will continue to be the gateway resource for many goods and services, including some that will be very difficult to do without. Still, there's no reason why distancing oneself from the tertiary economy has to be an all-or-nothing thing. Any step toward the direct production of goods and services for one's own use, with one's own labor, using resources under one's own direct control, is a step toward the world that will emerge after money; it's also a safety cushion against the disintegration of the tertiary economy going on around us.

AFTERWORD:

SMALL IS BEAUTIFUL

I

T HAS BEEN nearly four decades now since the limits to industrial civilization's trajectory of limitless material growth on
a limited planet have been clearly visible on the horizon of our
future. Over that time, a remarkable paradox has unfolded. The
closer we get to the limits to growth, the more clearly those limits impact our lives and the more obviously our current trajectory
points toward the brick wall of a difficult future, the less most people in the industrial world seem able to imagine any alternative to
driving the existing order of things until the wheels fall off.

This is as true in many corners of the activist community as it
is in the most unregenerate of corporate boardrooms. For most
of today's environmentalists, for example, renewable energy isn't
something that people ought to produce for themselves, unless
they happen to be wealthy enough to afford the rooftop PV systems that have become the latest status symbol in suburban neighborhoods on either coast in the US. It's something that utilities
and governments are supposed to bring online as fast as possible,

using existing power grids and habits of energy use, so that people in the industrial world can keep on using as much electricity as they want.

This approach is far from the only option available. In the energy crisis of the seventies, simple conservation and efficiency measures sent world petroleum consumption down by 15 percent in a single decade and caused comparable drops in other forms of energy consumption across the industrial world. Most of these measures went out the window in the final binge of the age of cheap oil that followed, so there's plenty of low-hanging fruit to pluck. That same era saw a great many thoughtful people envision ways that we could lead relatively comfortable and humane lives while consuming a great deal less energy and the products of energy than people in the industrial world do today.

It can be a troubling experience to turn the pages of *Rainbook* or *The Book of the New Alchemists*, to name only two of the better products of that mostly forgotten era, and compare the sweeping view of future possibilities that undergirded their approach to a future of energy and material shortages with the cramped imaginations of the present. It's even more troubling to notice that you can pick up yellowing copies of most of these books for a couple of dollars each in the used book trade, at a time when their practical advice is more relevant than ever, and their prophecies of what would happen if the road to sustainability was not taken are looking more prescient by the day. Thus our collective refusal to follow the lead of those who urged us to learn how to get by with less has not spared us the necessity of doing exactly that. That's the problem, ultimately, with driving headlong at a brick wall; you can stop by standing on the brake pedal or you can stop by hitting the wall, but either way, you're going to stop.

The brick wall ahead of us at this point in history marks, among other things, the conclusion of a long historical trajectory. From the early days of the industrial revolution into the early 1970s, the United States possessed the immense economic advantage of sizeable reserves of whatever the cutting-edge energy source happened to be. During what Lewis Mumford called the ecotechnic era,[1] when waterwheels were the prime mover for industry and canals were the core transportation technology, the United States prospered because it had an abundance of mill sites and internal waterways. During Mumford's paleotechnic era, when coal and railways replaced water and canal boats, the United States once again found itself blessed with huge coal reserves, and with the arrival of the neotechnic era, when petroleum and highways became the new foundation of power, the United States found that Nature had supplied it with so much oil that in 1950 it produced more petroleum than all other countries combined.

That trajectory came to an abrupt end in the 1970s, when nuclear power—expected by nearly everyone to be the next step in the sequence—turned out to be hopelessly uneconomical, and renewables proved unable to take up the slack at a cost the American political class was willing to pay. The neotechnic age, in effect, turned out to have no successor. Since then, the United States has been trying to stave off the inevitable sharp downward readjustment of our national standard of living and international importance following the depletion of the natural resources that once undergirded American economic and political power.

In the effort to prevent that readjustment from happening, the US has tried accelerating drawdown of natural resources, abandoning our national infrastructure, our industries and our agricultural hinterlands, and building ever more baroque systems of financial gimmickry to prop up our decaying economy with wealth from overseas. Over the last decade and a half, the government has resorted to systematically inflating speculative bubbles—and now, with its back to the wall, the US is printing money (under the euphemism of "quantitative easing") as though there's no tomorrow.[2] It's possible that the current US administration will be able to pull one more rabbit out of its hat and find a new gimmick to keep things going for a while longer. I have to confess that this does not look likely to me. Monetizing debt, as economists call the attempt to pay a nation's bills by means of the printing press, is a desperation move; it's hard to imagine any reason that it would have been chosen if there were any other option in sight.

What this means is that we may have just moved into the endgame of America's losing battle with the consequences of its own history. For many years now, people in the community of those concerned about the future have had, or thought they had, the luxury of ample time to make plans and take action. Every so often books would be written and speeches made claiming that something had to be done right away while there was still time, but most people took that as the rhetorical flourish it usually was, and went on with their lives in the confident expectation that the crisis was still a long ways off.

We may no longer have that option. If I read the signs correctly, America has finally reached the point where its economy is so far into ecological imbalance that disintegration is beginning in earnest. If so, a great many of the things most of us in this country have treated as permanent fixtures are likely to go away over the years immediately before us, as the United States transforms itself into a Third World country. The changes involved won't be sudden, and it seems unlikely that most of them will get much play in the domestic mass media; a decade from now, let's say, when half the American workforce has no steady work, decaying suburbs have mutated into squalid shantytowns and domestic insurgencies flare across the south and the mountain west, those who still have access to cable television will no doubt be able to watch talking heads explain how we're all better off than we were in 2000.

The economic debacle we are experiencing didn't just happen by chance; it became inevitable once the political classes of the industrial world convinced themselves that money was the source of

wealth rather than the mere measure of wealth it actually is. Decades of bad policy that encouraged making money at the expense
of the production of real wealth followed from those ideas. The
result was the transformation of a vast amount of paper "wealth" —
that is, money of one kind of another — into some malign equivalent of the twinkle dust of a children's fairy tale.

One way to make sense of the resulting collision between the
brittle front end of industrial civilization and the hard surface
of Nature's brick wall is to compare the spring of 2010 with the
autumn of 2007. Those two seasons had an interesting detail in
common. In both cases, the price of oil passed $80 a barrel af-
ter a prolonged period of price increases, and in both cases, this
was followed by a massive debt crisis. In 2007, largely driven by
speculation in the futures market, the price of oil kept on zooming
upwards, peaking just south of $150 a barrel before crashing back
to earth; so far, at least, there's no sign of a spike of that sort happening this time, although this is mostly because speculators are
focused on other assets these days.

In 2007, though, the debt crisis also resulted in a dramatic economic downturn, and as I write these words the world's chances of
dodging a repeat of the same experience do not look good. Here
in the US, most measures of general economic activity are faltering where they aren't plunging — the sole exceptions are those temporarily propped up by an unparalleled explosion of government
debt — and unemployment has become so deeply entrenched that
what to do about the very large number of Americans who have
exhausted the 99 weeks of unemployment benefits that current
law allows them in most states is becoming a significant political
issue. Even the illegal economy is taking a massive hit; a recent
news story broadcast on National Public Radio noted that the
price of marijuana has dropped so sharply that northern California, where it's a major cash crop, is seeing panic selling and sharp
economic contractions.[3]

What's going on here is precisely what *The Limits to Growth* warned about in 1973: the costs of continued growth have risen faster than growth itself, and are reaching a level that is forcing the economy to its knees. By "costs," of course, the authors of *The Limits to Growth* weren't talking about money. The costs that matter are energy, resources and labor; it takes a great deal more of all of these to extract oil from deepwater wells in the Gulf of Mexico or oil sands in Alberta, say, than it used to take to get sweet light crude from shallow strata in Pennsylvania or Texas, and since offshore drilling and oil sands make up an increasingly large share of what we've got left—those wells in Pennsylvania and Texas have been pumped dry, or nearly so—these real, nonmonetary costs have climbed steadily.

The price of oil in dollars functions here as a workable proxy, in the language of the tertiary economy, for the real cost of oil production in energy, resources and labor—the hard currency of the primary and secondary economies. The evidence of the last few years suggests that when the price of oil passes $80 a barrel, that's a sign that the primary and secondary costs of energy have risen high enough that the rest of the economy begins to crack under the strain. Since astronomical levels of debt have become standard practice all through today's global economy, the ability of marginal borrowers to service their debt is where the cracks show up first. In the fall of 2007, many of those marginal borrowers were homeowners in the US and UK; in the spring of 2010, they included entire nations.

Thus it's past time to realize that the age of abundance is over. The period from 1945 to 2005, when almost unimaginable amounts of cheap petroleum sloshed through the economies of the world's industrial nations, and transformed life in those nations almost beyond recognition, still shapes most of our thinking and nearly all of our expectations. Not one significant policy maker or mass media pundit in the industrial world has begun to talk about the impact of the end of the age of abundance. It's an open question if any of them have grasped how fundamental the changes will be as the new age of post-abundance economics begins to bear down.

Most ordinary people in the industrial world, for their part, are sleepwalking through one of history's great transitions. The issues that concern them are still defined entirely by the calculus of abundance. Most Americans these days, for example, worry about managing a comfortable retirement, paying for increasingly expensive medical care and providing their children with a college education and whatever amenities they consider important. It has not yet entered their darkest dreams that they need to worry about access to such basic necessities as food, clothing and shelter, the fate of local economies and communities shredded by decades of malign neglect, and the rise of serious threats to the survival of constitutional government and the rule of law.

Even among those who warn that today's Great Recession could bottom out at a level even worse than that reached in the Great Depression, very few have grappled with the consequences of a near-term future in which millions of Americans live in shanty-towns and struggle to find enough to eat every single day. To paraphrase Sinclair Lewis, that *did* happen here, only 80 years ago, and it did so at a time when the United States was a net exporter of industrial products and the world's largest producer and exporter of petroleum to boot. The same scale of economic decline in a nation that exports very little besides unpayable IOUs, and is the world's largest consumer and importer of petroleum, could easily have results much closer to those of the early twentieth century in Central Europe, to suggest only one example: near-universal impoverishment, food shortages, epidemics, civil wars and outbreaks of vicious ethnic cleansing, bracketed by two massive wars that each had body counts in the tens of millions.

It needs to be remembered, however, that this did not equate to the total collapse into a Cormac McCarthy future that so many people fantasize about these days. I've spent years wondering why it is that so many people seem unable to conceive of any future other than business as usual, on the one hand, and apocalyptic extremism on the other. Whatever the motives that drive this curious fixation, though, it results in a nearly complete blindness to the real risks the future is more likely to hold for us. It makes a useful exercise to take current survivalist notions about preparing for the future and ask yourself how many of them would have turned out to be useful over the decade or two ahead if someone had pursued exactly those strategies in Poland or Slovakia, let's say, in the years right before 1914.

Measure the difference between the real and terrible events of that period, on the one hand, and the fantasies of infinite progress or apocalyptic collapse that so often pass for realistic images of our future, on the other, and you have some sense of the gap that has to be crossed in order to make sense of a world after abundance. One way or another, we will cross that gap; the question is how many of us will make the crossing in advance, and have time to take constructive actions in response, and how many will do so only in retrospect, thinking ruefully of the dollars and hours spent preparing for an imaginary future while the real one was breathing down their necks.

I've talked at quite some length in this and other books about the kinds of preparations that will likely help individuals, families and communities deal with the future of resource shortages, economic implosion, political breakdown and potential civil war that the missed opportunities and purblind decisions of the last 30 years have made agonizingly likely here in the United States and, with an infinity of local variations, elsewhere in the industrial world. Those points remain crucial. It still makes a great deal of sense to start growing some of your own food, to radically downscale your dependence on complex technological systems, to reduce your energy consumption as far as possible, to free up at least one family member from the money economy for full-time work in the domestic economy and so on.

Still, there's another dimension to all this and it has to be mentioned, though it's certain to raise hackles. For the last three centuries, and especially for the last half century or so, it's become increasingly common to define a good life as one provided with the largest possible selection of material goods and services. That definition has become so completely hardwired into our modern ways of thinking that it can be very hard to see past it. Of course there are certain very basic material needs without which a good life is impossible, but those are a good deal fewer and simpler than contemporary attitudes assume, and once those are provided, material abundance becomes a much more ambivalent blessing than we like to think.

In a very real sense, this way of thinking mirrors the old joke about the small boy with a hammer who thinks everything is a nail. In an age of unparalleled material abundance, the easy solution for any problem or predicament was to throw material wealth at it. That did solve some problems, but it arguably worsened others, and left the basic predicaments of human existence untouched. Did it really benefit anyone to spend trillions of dollars and the talents of some of our civilization's brightest minds creating high-end medical treatments to keep the very sick alive and miserable for a few extra months of life, for example, so that we could pretend to ourselves that we had evaded the basic human predicament of the inevitability of death?

Whatever the answer, the end of the age of abundance draws a line under that experiment. Within not too many years, it's safe to predict, only the relatively rich will have the dubious privilege of spending the last months of their lives hooked up to complicated life support equipment. The rest of us will end our lives the way our great-grandparents did: at home, more often than not, with family members or maybe a nurse to provide palliative care while our bodies do what they were born to do and shut down. Within not too many years, more broadly, only a very few people anywhere in the world will have the option of trying to escape the core uncertainties and challenges of human existence by chasing round after round of consumer goods; the rest of us will count ourselves lucky to have our basic material needs securely provided for, and will have to deal once again with fundamental questions of meaning and value.

It will probably come as no surprise that here again E. F. Schumacher was well ahead of the rest of us. His final, posthumously published book, *A Guide For The Perplexed*, left behind the economic topics that occupied most of his public career to ask hard questions, and propose provocative answers, concerning the most basic issues surrounding human life. It would take another book the size of this one, if not larger, to offer any meaningful comment on the points Schumacher raised in this slender volume, but one theme that runs through it is a recognition that material wealth, however defined, only satisfies a certain sharply limited number of human needs, and a life oriented entirely toward material accumulation has failed to deal with most of the possibilities open to human existence.

In the recognition of that failure, other possibilities open up. Still, the habits of thought instilled in us by an age of abundance get in the way of recognizing most of those possibilities. It's worth noting that while there's been plenty of talk about the monasteries of the Dark Ages among people who are aware of the impending decline and fall of our civilization, next to none of it has discussed, much less dealt with, the secret behind the success of monasticism: the deliberate acceptance of extreme material poverty. Quite the contrary: all the plans for lifeboat ecovillages I've encountered so far, at least, aim at preserving some semblance of a middle-class lifestyle into the indefinite future. That choice puts these projects in the same category as the lavish villas in which the wealthy inhabitants of Roman Britain hoped to ride out their own trajectory of decline and fall—a category mostly notable for its long history of total failure.

The European Christian monasteries that preserved Roman culture through the Dark Ages did not offer anyone a middle-class lifestyle by the standards of their own time, much less those of ours. Neither did the Buddhist monasteries that preserved Heian culture through the Sengoku Jidai, Japan's bitter age of wars, or the Buddhist and Taoist monasteries that preserved classical Chinese culture through a good half-dozen cycles of collapse, or any of the other examples of successful monasticism in history. Monasteries in all these cases were places people went to be very, very poor. That was the secret of their achievements, because when you reduce your material needs to the absolute minimum, the energy you don't spend maintaining your standard of living can be put to work doing something more useful.

Now it's probably too much to hope for that some similar movement might spring into being here and now; we're a couple of centuries too soon for that. The great age of Christian monasticism in the West didn't begin until the sixth century CE, by which time the Roman economy of abundance had been gone for so long that nobody even pretended that material wealth was an answer to the human condition. Still, the monastic revolution kick-started by Benedict of Nursia drew on a long history of Christian monastic ventures; those unfolded in turn from the first tentative communal hermitages of early Christian Egypt; and all these projects, though this is not often mentioned, took part of their inspiration and a good deal of their ethos from the Stoics of Pagan Greece and Rome.

Movements away from an obsession with material wealth are in fact very common in civilizations that have passed the Hubbert peak of their own core resource base. There's good reason for that. In a contracting economy, it becomes easier to notice that the less you need, the less vulnerable you are to the ups and downs of fortune, and the more you can get done of whatever it is that you happen to want to do. That's an uncongenial lesson at the best of times, and during times of material abundance you won't find many people learning it; it's one of the marks of Schumacher's insight that he did learn it, and tried in various ways to communicate it to a mostly uninterested world. Still, in the world after abundance, it's hard to think of a lesson that deserves more careful attention.

Notes

Introduction: A Guide for the Perplexed

1. Strictly speaking, this is not actually a Nobel prize, as it was founded by the Swedish government rather than by Alfred Nobel, who created the others.

2. E. F. Schumacher, *Small is Beautiful: Economics as if People Mattered*, Harper, 1973, p. 9.

3. This point is developed in detail in John Michael Greer, *The Long Descent*, New Society Publishers, 2008 and *The Ecotechnic Future*, New Society Publishers, 2009.

Chapter One: The Failure of Economics

1. See, among other accounts, that in Carmen M. Reinhart and Kenneth S. Rogoff, *This Time Is Different: Eight Centuries of Financial Folly*, Princeton University Press, 2009.

2. His predictions, and those of many others, are archived at housingpanic.blogspot.com.

3. Noriel Roubini and Stephen Mihm, *Crisis Economics: A Crash Course on the Future of Finance*, Penguin, 2010.

4. John Kenneth Galbraith, *A Short History of Financial Euphoria*, Penguin, 1994.

5. Roger Lowenstein, *When Genius Failed: The Rise and Fall of Long-Term Capital Management*, Random House, 2000.

6. David A. Moss, *A Concise Guide to Macroeconomics*, Harvard Business School Press, 2007, p. 141.

7. See particularly Steve Mohr, "Projection of World Fossil Fuel Production with Supply and Demand Interactions," Ph.D. thesis, University of Newcastle, Australia, February 2010.

8. Glenn Sweetnam, "Meeting the World's Demand for Liquid Fuels," presentation at Energy Information Agency's 2009 Energy Conference, Washington, DC, p. 8.

9. Kurt Cobb, "Faith-based economics in two graphs," energybulletin .net/52294.

10. Adam Smith, *The Wealth of Nations*, William Benton, 1952, p. 55.

11. Garrett Hardin, "The Tragedy of the Commons," *Science*, Vol. 162, no. 3859 (1968), pp. 1243–1248.

12. See, for example, Elinor Ostrom, *Governing the Commons: The Evolution of Institutions for Collective Action*, Cambridge University Press, 1990.

13. Smith, *Wealth of Nations*, pp. 51–62.

14. Schumacher, *Small is Beautiful*, pp. 16–22.

15. The relation between thermodynamics and economics has been usefully explored in Nicholas Georgescu-Roegen, "Myths about Energy and Matter," *Growth and Changes*, Vol. 10, issue 1, January 1979, pp. 16–23.

16. Bertram Gross, *Friendly Fascism: The New Face of Power in America*, South End Press, 1980.

17. shadowstats.com provides a full range of standard US economic statistics calculated according to the methods used up until the 1970s. A comparison between those statistics and those issued by the US government, based on the identical set of raw data, is an eye-opening experience.

18. Moss, *Concise Guide*, p. 21.

Chapter Two: The Three Economies

1. For an example, see Moss, *Concise Guide*, 2007.

2. David Ricardo, *The Principles of Political Economy and Taxation*, Richard D. Irwin, 1963. The phrase "original and indestructible" is from p. 29.

3. Karl Marx, *Capital*, trans. Samuel Moore and Edward Aveling, William Benton, 1952, p. 16.

4. For an example see Edwin G. Dolan, *TANSTAAFL: The Economic Strategy for Environmental Crisis*, Holt, Rimehart and Winston, 1971.

5. A readable summary of soil science can be found in Firman E. Bear, *Earth: The Stuff of Life*, University of Oklahoma Press, 1986.

6. See F. H. King, *Farmers of Forty Centuries*, Rodale, 1973.

7. R. Costanza et al., "The value of the world's ecosystem services and natural capital," *Nature* 387 (1997), pp. 253–260.

8. For a good summary see Clive Ponting, *A Green History of the World: The Environment and the Collapse of Great Civilizations*, St. Martins, 1992.

9. Schumacher, *Small is Beautiful*, pp. 49–52.

10. See the discussion in Alf Hornborg, *The Power of the Machine: Global Inequalities of Economy, Technology, and the Environment*, Alta Mira Press, 2001.

11. A useful summary can be found in Reinhart and Rogoff, *This Time Is Different*.

12. I have discussed this point in more detail in Chapter 3 of *The Ecotechnic Future*.

Chapter Three: The Metaphysics of Money

1. Smith, *Wealth of Nations*, p. 182.

2. Ilya Prigogine and Isabelle Stengers, *Order Out of Chaos: Man's New Dialogue with Nature*, Bantam, 1984.

3. John Kenneth Galbraith, *The Great Crash 1929*, Houghton Mifflin, 1954, pp. 144–145.

Chapter Four: The Cost of Energy

1. Schumacher, *Small is Beautiful*, p. 123.

2. Sir Arthur Stanley Eddington, *The Nature of the Physical World*, Macmillan, 1928, p. 74.

3. Garrett Hardin, *Filters Against Folly: How to Survive Despite Economists, Ecologists, and the Merely Eloquent*, Penguin, 1985, p. 30.

4. The term is from Thomas Friedman, *Hot, Flat, and Crowded: Why We Need a Green Revolution—And How It Can Renew America*, Farrar, Straus and Giroux, 2008.

5. See Jeremy K. Leggett, *The Empty Tank: Oil, Gas, Hot Air, and the Coming Global Financial Catastrophe*, Random House, 2005, p. 67.

6. Ken Butti and John Perlin, *A Golden Thread: 2500 Years of Solar Architecture and Technology*, Cheshire Books, 1980, pp. 106–9.

7. Ibid, p. 101.

8. Ibid, pp. 63–75.

9. Ibid, p. 72.

Chapter Five: The Appropriate Tools

1. Richard C. Duncan, "The life-expectancy of industrial civilization: the decline to global equilibrium," *Population and Environment*, 14:4 (1993), pp. 325–357.

2. The classic account is Jared Diamond, *Collapse: How Societies Choose to Fail or Succeed*, Penguin, 2005.

3. See, for example, Hannes Kunz and Stephen Balogh, "Fake firemen — why are we cheating ourselves on energy?" International Institute for Energy Resources, wiier.ch.

4. See Schumacher, *Small is Beautiful*, pp. 163–220.

5. The experience of the former Soviet Union, which is experiencing a population contraction sufficient to reduce its population by half before 2100, is a case in point. See, for example, Dmitry Orlov, *Reinventing Collapse: The Soviet Example and American Prospects*, New Society Publishers, 2008.

6. Arnold Toynbee, *A Study of History, Volume V: The Disintegrations of Civilizations*, Oxford University Press, 1939, pp. 58–193.

7. See Charles A.S. Hall and John W. Day, Jr., "Revisiting the Limits to Growth After Peak Oil," *American Scientist*, Vol. 97 (May–June 2009), pp. 230–237.

Chapter Six: The Road Ahead

1. Galbraith, *A Short History*.

2. Hardin, "Tragedy of the Commons."

3. See, for example, Dolan, *TANSTAAFL*.

4. Arthur B. Laffer, Stephen Moore and Peter J. Tanous, *The End of Prosperity: How Higher Taxes Will Doom the Economy—If We Let It Happen*, Threshold Editions, 2008.

5. Schumacher, *Small is Beautiful*, pp. 272–292.

6. David Korten, *When Corporations Rule the World*, Berrett-Koehler, 2001.

7. Discussed in Schumacher, *Small is Beautiful*, pp. 244–246.

8. Brian Ward-Perkins, *The Fall of Rome and the End of Civilization*, Oxford University Press, 2005.

9. A good recent source is Steven Epstein, *Wage Labor and Guilds in Medieval Europe*, University of North Carolina Press, 1991.

10. Warren Johnson, *Muddling Toward Frugality: A New Social Logic for a Sustainable World*, rev. ed., Easton Studio Press, 2010.

Afterword: Small Is Beautiful

1. Lewis Mumford, *Technics and Civilization*, Harcourt Brace, 1934.

2. See the discussion in Tim Morgan, *Dangerous Exponentials: A Radical Take on the Future*, Tullett Prebon, 2010, pp. 16–19.

3. Michael Montgomery, "Plummeting marijuana prices create a panic in California," National Public Radio, May 15, 2010. Archived at npr.org/templates/story/story.php?storyid=126806429.

Bibliography

Asafu-Adjaye, John, *Environmental Economics for Non-Economists*, World Scientific, 2000.

Bear, Firman E., *Earth: The Stuff of Life*, University of Oklahoma Press, 1986.

Brandt, Barbara, *Whole Life Economics*, New Society Publishers, 1995.

Butti, Ken, and John Perlin, *A Golden Thread: 2500 Years of Solar Architecture and Technology*, Cheshire Books, 1980.

Chambers, Nicky, Craig Simmons and Mathis Wackernagel, *Sharing Nature's Interest: Ecological Footprints as an Indicator of Sustainability*, Earthscan, 2000.

Cobb, Kurt, "Faith-based economics in two graphs," energybulletin .net/52294. Cited April 4, 2010.

Costanza, R., R. d'Arge, R. de Groot, S. Farber, M. Grasso, B. Hannon, K. Limburg, S. Naeem, R. O'Neill, J. Paruelo, R. Raskin, P. Sutton and M. van den Belt, "The value of the world's ecosystem services and natural capital," *Nature*, 387, 1997.

Diamond, Jared, *Collapse: How Societies Choose to Fail or Succeed*, Penguin, 2005.

Dolan, Edwin G., *TANSTAAFL: The Economic Strategy for Environmental Crisis*, Holt, Rinehart and Winston, 1971.

Duncan, Richard C., "The life-expectancy of industrial civilization: the decline to global equilibrium," *Population and Environment*, 14:4, 1993.

Eddington, Sir Arthur Stanley, *The Nature of the Physical World*, Macmillan, 1928.

Ekins, Paul, *The Gaia Atlas of Green Economics*, Doubleday, 1992.

Epstein, Steven A., *Wage Labor and Guilds in Medieval Europe*, University of North Carolina Press, 1991.

Friedman, Thomas, *Hot, Flat, and Crowded: Why We Need a Green Revolution—And How It Can Renew America*, Farrar, Straus and Giroux, 2008.

Froggatt, Anthony, *Sustainable Energy Security: Strategic Risks and Opportunities for Business*, Chatham House, 2010.

Galbraith, John Kenneth, *A Short History of Financial Euphoria*, Penguin, 1994.

———, *The Great Crash 1929*, Houghton Mifflin, 1954.

Georgescu-Roegen, Nicholas, "Myths about Energy and Matter," *Growth and Change*, 10, 1979.

———, *The Entropy Law and the Economic Process*, Harvard University Press, 1971.

———, "The Entropy Law and the Economic Process in Retrospect," *Eastern Economic Journal*, Vol. 12, No. 1, January–March 1980.

Gimpel, Jean, *Medieval Machine: The Industrial Revolution of the Middle Ages*, Penguin, 1977.

Gorman, Tom, *The Complete Idiot's Guide to Economics*, Alpha Books, 2003.

Greer, John Michael, *The Ecotechnic Future*, New Society Publishers, 2009.

———, *The Long Descent*, New Society Publishers, 2008.

Gross, Bertram, *Friendly Fascism: The New Face of Power in America*, South End Press, 1980.

Hall, Charles A.S., and John W. Day, Jr., "Revisiting the Limits to Growth After Peak Oil," *American Scientist*, Vol. 97, May–June 2009.

Hardin, Garrett, *Filters Against Folly: How to Survive Despite Economists, Ecologists, and the Merely Eloquent*, Penguin Books, 1985.

———, "The Tragedy of the Commons," *Science*, Vol. 162, No. 3859, 1968.

Hornborg, Alf, *The Power of the Machine: Global Inequalities of Economy, Technology, and Environment*, Alta Mira Press, 2001.

Johnson, Warren, *Muddling Toward Frugality: A New Social Logic for a Sustainable World*, rev. ed., Easton Studio Press, 2010.

King. F. H., *Farmers of Forty Centuries*, Rodale Press, 1973.

Korten, David, *When Corporations Rule the World*, Berrett-Koehler, 2001.

Kuhn, Thomas S., *The Structure of Scientific Revolutions*, University of Chicago Press, 1996.

Kunz, Hannes, and Stephen Balogh, "Fake firemen—why are we cheating ourselves on energy?," International Institute for Energy Resources, iier.ch. Cited June 26, 2010.

Laffer, Arthur B., Stephen Moore, and Peter J. Tanous, *The End of Prosperity: How Higher Taxes Will Doom the Economy—If We Let It Happen*, Threshold Editions, 2008.

Leggett, Jeremy K., *The Empty Tank: Oil, Gas, Hot Air, and the Coming Global Financial Catastrophe*, Random House, 2005.

Lewis, Michael, *The Big Short: Inside the Doomsday Machine*, W. W. Norton, 2010.

Lowenstein, Roger, *When Genius Failed: The Rise and Fall of Long-Term Capital Management*, Random House, 2000.

Marx, Karl, *Capital*, tr. Samuel Moore and Edward Aveling, William Benton, 1952.

Mohr, Steve, "Projection of World Fossil Fuel Production with Supply and Demand Interactions," Ph.D. thesis, University of Newcastle, Australia, February 2010.

Montgomery, Michael, "Plummeting marijuana prices create a panic in California," National Public Radio, May 15, 2010; archived at npr.org/templates/story/story.php?storyid=126806429.

Morgan, Dr. Tim, *Dangerous Exponentials: A Radical Take on the Future*, Tullett Prebon, 2010.

Moss, David A., *A Concise Guide to Macroeconomics*, Harvard Business School Press, 2007.

Mumford, Lewis, *Technics and Civilization*, Harcourt Brace and Co., 1934.

Orlov, Dmitry, *Reinventing Collapse: The Soviet Example and American Prospects*, New Society Publishers, 2008.

Ostrom, Elinor, *Governing the Commons: The Evolution of Institutions for Collective Action*, Cambridge University Press, 1990.

Ponting, Clive, *A Green History of the World: The Environment and the Collapse of Great Civilizations*, St. Martins, 1992.

Prigogine, Ilya, and Isabelle Stengers, *Order Out Of Chaos: Man's New Dialogue with Nature*, Bantam, 1984.

Prugh, Thomas, et al., *Natural Capital and Human Economic Survival*, ISEE Press, 1995.

Redclift, Michael, *Wasted: Counting the Costs of Global Consumption*, Earthscan Publications, 1996.

Reinhart, Carmen M., and Kenneth S. Rogoff, *This Time Is Different: Eight Centuries of Financial Folly*, Princeton University Press, 2009.

Ricardo, David, *The Principles of Political Economy and Taxation*, Richard D. Irwin, 1963.

Roubini, Nouriel, and Stephen Mihm, *Crisis Economics: A Crash Course on the Future of Finance*, Penguin, 2010.

Schumacher, E. F., *A Guide for the Perplexed*, Abacus, 1977.

———, *Small is Beautiful: Economics as if People Mattered*, Harper, 1973.

Shea, Robert, and Robert Anton Wilson, *Illuminatus!*, Dell, 1975.

Smith, Adam, *The Wealth of Nations*, William Benton, 1952.

Sweetnam, Glenn, "Meeting the World's Demand for Liquid Fuels," presentation at Energy Information Agency's 2009 Energy Conference, Washington, DC.

Tainter, Joseph, *The Collapse of Complex Socieities*, Cambridge University Press, 1990.

Toynbee, Arnold, *A Study of History, Vol. V: The Disintegrations of Civilizations*, Oxford University Press, 1939.

US Energy Information Administration, *Annual Energy Outlook 2009*, Government Printing Office, 2009.

Ward-Perkins, Brian, *The Fall of Rome and the End of Civilization*, Oxford University Press, 2005.

Index

hippogriff(s), 26
honeybees, 60, 79-80
Hopkins, Rob, 228
household economy, 64, 65, 225

I

Imaginaria, 174-175
inflation, 117, 146-147
intermediate technology, 4, 141, 174
internet, 155-160, 163-4
investment, 109-113

J

Jackson, Michael, 34
Johnson, Lyndon B., 34
Johnson, Warren, 223-224

K

Keynes, John Maynard, 4
King, F.H., 177
Kuhn, Thomas, 55-59

L

Lehrer, Tom, 172
libraries, public, 162-163
Liebig's law of the minimum, 37, 59
"lifeboat communities," 189-190
Limits to Growth, The, 188, 238
Long Term Capital Management (LTCM), 14-15, 18

M

Mackay, Rev. Charles, 12
Marx, Karl, 50, 195

Printed in the USA
CPSIA information can be obtained
at www.ICGtesting.com
JSHW011717251024
72426JS00004B/41